Quantum Computer Systems

Research for Noisy Intermediate-Scale Quantum Computers

Synthesis Lectures on Computer Architecture

Editors
Natalie Enright Jerger, *University of Toronto*
Margaret Martonosi, *Princeton University*

Founding Editor Emeritus
Mark D. Hill, *University of Wisconsin, Madison*

Synthesis Lectures on Computer Architecture publishes 50- to 100-page publications on topics pertaining to the science and art of designing, analyzing, selecting and interconnecting hardware components to create computers that meet functional, performance and cost goals. The scope will largely follow the purview of premier computer architecture conferences, such as ISCA, HPCA, MICRO, and ASPLOS.

Transactional Memory
James R. Larus and Ravi Rajwar
2006

Quantum Computing for Computer Architects
Tzvetan S. Metodi and Frederic T. Chong
2006

Quantum Computer Systems: Research for Noisy Intermediate-Scale Quantum Computers
Yongshan Ding and Frederic T. Chong

ISBN: 978-3-031-00637-1 paperback
ISBN: 978-3-031-01765-0 ebook
ISBN: 978-3-031-00062-1 hardcover

DOI 10.1007/978-3-031-01765-0

A Publication in the Springer series
SYNTHESIS LECTURES ON ADVANCES IN AUTOMOTIVE TECHNOLOGY

Lecture #51
Series Editors: Natalie Enright Jerger, *University of Toronto*
                Margaret Martonosi, *Princeton University*
Founding Editor Emeritus: Mark D. Hill, *University of Wisconsin, Madison*
Series ISSN
Print 1935-3235   Electronic 1935-3243

Quantum Computer Systems

Research for Noisy Intermediate-Scale
Quantum Computers

Yongshan Ding
University of Chicago

Frederic T. Chong
University of Chicago

SYNTHESIS LECTURES ON COMPUTER ARCHITECTURE #51

ABSTRACT

This book targets computer scientists and engineers who are familiar with concepts in classical computer systems but are curious to learn the general architecture of quantum computing systems. It gives a concise presentation of this new paradigm of computing from a computer systems' point of view without assuming any background in quantum mechanics. As such, it is divided into two parts. The first part of the book provides a gentle overview on the fundamental principles of the quantum theory and their implications for computing. The second part is devoted to state-of-the-art research in designing practical quantum programs, building a scalable software systems stack, and controlling quantum hardware components. Most chapters end with a summary and an outlook for future directions. This book celebrates the remarkable progress that scientists across disciplines have made in the past decades and reveals what roles computer scientists and engineers can play to enable practical-scale quantum computing.

KEYWORDS

quantum computing, computer architecture, quantum compilation, quantum programming languages, quantum algorithms, noise mitigation, error correction, qubit implementations, classical simulation

Contents

Preface

Quantum computing is at a historic time in its development and there is a great need for research in quantum computer systems. This book stems from a course we co-taught in 2018 and the research efforts of the EPiQC NSF Expedition in Computing and others. Our goal is to provide a broad overview of some of the emerging research areas in the development of practical computing systems based upon emerging noisy intermediate-scale quantum hardware. It is our hope that this book will encourage other researchers in the computer systems community to pursue some of these directions and help accelerate real-world applications of quantum computing.

Despite the impressive capability of today's digital computers, there are still some computational tasks that are beyond their reach. Remarkably, some of those tasks seem to be relatively easy with a quantum computer. Over the past four decades or so, our understanding of the theoretical power of quantum and skills in quantum engineering has advanced significantly. Small-scale prototypes of progammable quantum computers are emerging from academic and industry labs around the world. This is undoubtedly an exciting time, as we may be soon fortunate enough to be among the first to witness the application of quantum computers on problems that are unfeasible for today's classical computers. What has been truly remarkable is that the field of quantum information science has brought scientists together across disciplines—physicists, electrical engineers, computer architects, and theorists, just to name a few.

Looking back at the historical progress in digital computers, we remark upon the three major milestones that led to the integration of millions of computational units that make up the computing power in today's computers: low-cost integrated circuit technology, efficient architectural design, and interconnected software ecosystem. It is not too unrealistic to assume that the evolution of quantum computers will follow a similar trajectory; we are starting to see some innovations in hardware, software, and architecture designs that have the potential to scale up well. The progress and prospect of the new paradigm of computing has motivated us to write this Synthesis Lecture, which hopefully can bring together more and more computer scientists and engineers to join the expedition to practical-scale quantum computation.

This introduction to quantum computer systems should primarily appeal to computer systems researchers, software engineers, and electrical engineers. The focus of this book is on systems research for noisy intermediate-scale quantum (NISQ) computers, highlighting the recent progress and addressing the near-term challenges for realizing the computational power of QC systems.

Reading This Book

The aim of this book is to provide computer systems researchers and engineers with an introductory guide to the general principles and challenges in designing practical quantum computing systems. Compared to its predecessor in the series, *Quantum Computing for Computer Architects* by Metodi, Faruque, and Chong [1], this book targets near-term progress and prospects of quantum computing. Throughout the book, we emphasize how computer systems researchers can contributes to the exciting emerging field. As such, the structure of this book is as follows. Chapter 2 reviews the central concepts in quantum computation, compares and contrasts with those of classical computation, and discusses the leading technologies for implementing qubits. Chapter 3 summarizes the general features in quantum algorithms and reviews some of the important NISQ applications.

The second part of the book starts in Chapter 4 with an overview of the quantum architectural vertical stack and the cross-cutting themes that enable synergy among the different disciplines in the field. The rest of the book illuminates the opportunities in quantum computer systems research, broadly split into five tracks: (i) Chapter 5 describes existing quantum programming languages and techniques for debugging and verification; (ii) Chapter 6 introduces important quantum compilation methods including circuit optimization and synthesis; (iii) Chapter 7 dives into low-level quantum controls, pulse generation, and calibration; (iv) a number of noise mitigation and error correction techniques are reviewed in Chapter 8; (v) Chapter 9 discusses different methods in classical simulations of quantum circuits and their implications; and (vi) a summary of progress and prospects of quantum computer systems research can be found in Chapter 10.

The reader is encouraged to start with the Summary and Outlook section in some chapters for a quick overview of fundamental concepts, highlights of state-of-the-art research, and discussions of future directions.

Yongshan Ding and Frederic T. Chong
Chicago, June 2020

Acknowledgments

Our views in the book are strongly informed by ideas formed from discussions with Yuri Alexeev, Kenneth Brown, Chris Chamberland, Isaac Chuang, Andrew Cross, Bill Fefferman, Diana Franklin, Alexey Gorshkov, Hartmut Haeffner, Danielle Harlow, Aram Harrow, Henry Hoffman, Andrew Houck, Ali Javadi-Abhari, Jungsang Kim, Peter J. Love, Margaret Martonosi, Akimasa Miyake, Chris Monroe, William Oliver, John Reppy, David Schuster, Peter Shor, Martin Suchara, members of the EPiQC Project (Enabling Practical-scale Quantum Computation, an NSF Expedition in Computing), and members of the STAQ Project (Software-Tailored Architecture for Quantum co-design). Thanks are extended to the students who took the 2018 course on quantum computer systems for their helpful lecture scribing notes: Anil Bilgin, Xiaofeng Dong, Shankar G. Menon, Jean Salac, and Lefan Zhang, among others.

Thanks to Morgan & Claypool Publishers for making the publication this book possible. Many thanks to Michael Morgan, who invited us to write on the subject, for his patience and encouragement. Thanks also to our Synthesis Lecture series editors Natalie Enright Jerger and Margaret Martonosi, who shepherded this project to its final product. YD and FTC are grateful to Frank Mueller and the anonymous reviewers for providing in-depth comments and suggestions on the original manuscript. Thanks to Sara Kreiman for her thorough copyedit of the book.

YD has learned a tremendous amount from his advisor FTC, and is very grateful for FTC's mentorship in quantum information science research and education. YD also thanks Ryan O'Donnell, who first introduced him to the field of quantum computation and information. YD worked on this book while visiting the Massachusetts Institute of Technology. YD especially thanks Isaac Chuang, Aram Harrow, and Peter Shor for the many inspiring discussions during his visit. YD thanks all of his colleagues, friends, and relatives for their encouragement and support in writing and finishing the book, especially Meizi Liu, and YD's parents, Genlin Ding and Shuowen Feng.

Finally, YD and FTC gratefully acknowledge the support from the National Science Foundation, specifically by EPiQC, an NSF Expedition in Computing, under grants CCF-1730449, in part by STAQ, under grant NSF Phy-1818914, and in part by DOE grants DE-SC0020289 and DE-SC0020331.

Yongshan Ding and Frederic T. Chong
Chicago, June 2020

List of Notations

The nomenclature and notations used in this book may be unfamiliar to many readers and may have different meanings in a different context. We devote this section to clarifying some of the conventions this book uses to prevent confusion.

Systems Terminology

- *Adiabatic quantum computing* is a model of analog quantum computing where a quantum system remains in the ground state energy.

- *Analog quantum computing (AQC)* is a model of quantum computation such that the state of a quantum system is evolved smoothly.

- *Boolean circuit* is a model of classical computation that expresses computation by sending data through a combination of logic gates.

- *FT* refers to being fault tolerant; a fault-tolerant quantum computer relies on quantum error correction.

- The *gate scheduling* problem is to design an ordering or synchronization of quantum gates to be applied to the qubits in the target architecture, under constraints such as data dependencies, parallelism, communication, and noise.

- *Hamiltonian* refers to the mathematical representation of the energy configuration of a physical system. It is commonly used as a linear algebraic operator in quantum mechanics.

- *Host processor* is an abstraction that refers to the classical computer that controls the processes in quantum computer systems.

- *Quantum annealing* is a model of analog quantum computing where-in the quantum systems interact with the thermal environment.

- *Lambda calculus* is a model of classical computation based on functional expressions using variable binding and substitution.

- *Measurement-based quantum computing* (MBQC) is a model of computation that performs computation via only measurements on qubits previously initialized to a cluster state.

- A *NISQ* computer refers to a noisy intermediate-scale quantum computer.

- *Turing machine* is a model of classical computation for abstract computing machines based on manipulating data sequentially on a strip of tape following a set of rules.

- *Quantum compiling* refers to the framework for efficiently implementing a given quantum program or target unitary to high precision, using gates from a set of primitive instructions supported in the underlying quantum architecture.

- *Quantum communication* is a branch of quantum technology where-in entangled qubits are used to encrypt and transmit data.

- *Quantum circuit synthesis* refers to the technique that constructs a gate out of a series of primitive operation.

- *Quantum device topology* (or device connectivity) describes the layout of the physical qubits and the allowed direct interactions between any pair of qubits.

- *Quantum logic gates* (or *qubit operations* or *quantum instructions*) are transformations to be applied to qubits, represented by unitary matrices.

- The *qubit mapping* problem aims to find an optimal mapping from the qubit registers in a quantum program to the qubits in the target architecture, under constraints such as system size, data dependencies, communication, ancilla reuse, and noise.

- *Quantum processing unit (QPU)* refers to a hardware component that implements qubits as well as the control apparatus.

- A *quantum program* is an abstraction that refers to the sequence of instructions and control flow that a quantum computer must follow according to a protocol or an algorithm.

- *Quantum sensing* is a branch of quantum technology that takes advantage of quantum coherence to perform measurements of physical quantities.

- *Quantum simulation* is a branch of quantum technology that studies the structures and properties of electronic or molecular systems.

- *Schoelkopf's law* is an empirical scaling projection for quantum decoherence—delayed by a factor of 10 roughly every three years.

- The *von Neumann architecture* is a stored-program computer architecture that controls instruction fetch and data operations via a common system computer bus.

Linear Algebra and Probability in Quantum Computing

- The *basis* of a qubit is a set of linearly independent vectors that span the Hilbert space. The two most common bases for single qubits are the computational basis (z basis):

$$\{|0\rangle, |1\rangle\} \equiv \left\{ \begin{pmatrix} 1 \\ 0 \end{pmatrix}, \begin{pmatrix} 0 \\ 1 \end{pmatrix} \right\},$$

and the Fourier basis (x basis):

$$\{|+\rangle, |-\rangle\} \equiv \left\{ \begin{pmatrix} 1/\sqrt{2} \\ 1/\sqrt{2} \end{pmatrix}, \begin{pmatrix} 1/\sqrt{2} \\ -1/\sqrt{2} \end{pmatrix} \right\}.$$

- The *Bloch sphere* is a visualization of single-qubit Hilbert space $\mathcal{H}$ in three-dimensional Euclidean space $\mathbb{R}^3$:

$$\rho(x, y, z) = \frac{1}{2}(I + x\sigma_x + y\sigma_y + z\sigma_z).$$

- The *bra vector* is the conjugate transpose of a ket vector:

$$\langle\psi| = \begin{pmatrix} \alpha^* & \beta^* \end{pmatrix}.$$

- A *cluster state* is a quantum state defined by a graph, where the nodes in the graph are qubits initialized to $|+\rangle$ state, and the edges are controlled-Z gates between the qubits.

- A *complex number* $z \in \mathbb{C}$ is a number in the form of $a + bi$, where a, b are real numbers and i is an imaginary unit satisfying $i^2 = -1$. a is called the read part, and b is called the imaginary part of z. The *conjugate* of z is $z^* = a - bi$.

- The *conjugate transpose* of a matrix M is denoted as $M^\dagger$ whose matrix elements are:

$$[M^\dagger]_{ij} = [M]_j i^*.$$

- An *EPR* pair refers to two qubits in the quantum state $|epr\rangle = (|00\rangle + |11\rangle)/\sqrt{2}$.

- The common mapping is $|0\rangle$ for *ground energy state*, and $|1\rangle$ for *first excited energy state*. In the context of the physical implementation of a qubit, the computational basis corresponds to the discrete energy levels.

- A complex square matrix is *Hermitian* if its complex conjugate transpose $H^\dagger$ is equal to itself:

$$H^\dagger = H.$$

- The *Hilbert space* $\mathcal{H}$ is complex inner product space in which a n-qubit quantum state is a 2^n-dimensional vector of complex entries.

- The *inner product* of two quantum states $|\psi\rangle = \sum_j \alpha_j |j\rangle$, $|\phi\rangle = \sum_k \beta_k |k\rangle$ is $\langle\psi|\phi\rangle = \sum_i = \alpha_i^* \beta_i$.

- An *identity matrix* I is a matrix with 1 along the diagonal and 0 everywhere else.

- For any real number $p \geq 1$, the ℓ_p *norm* of a vector $\mathbf{x} = (x_1, \ldots, x_n)$ is defined as

$$||\mathbf{x}||_p = \left(\sum_{i=1}^{n} |x_i|^p \right)^{1/p}.$$

- e^M and $\exp(M)$ are notations for *matrix exponential*, which is defined as:

$$e^M = \sum_{k=0}^{\infty} \frac{1}{k!} M^k.$$

- A *mixed quantum state* or *density matrix* is a probability ensemble of pure quantum states: $\rho = \sum_i p_i |\psi_i\rangle \langle\psi_i|$.

- The *Pauli matrices* are

$$\sigma_x = \begin{pmatrix} 0 & 1 \\ 1 & 0 \end{pmatrix}, \sigma_y = \begin{pmatrix} 0 & -i \\ i & 0 \end{pmatrix}, \sigma_z = \begin{pmatrix} 1 & 0 \\ 0 & -1 \end{pmatrix}.$$

- A *probability distribution* refers to a finite set of non-negative real numbers p_i that sums to 1: $p_i \geq 0$ and $\sum_i p_i = 1$.

- A *quantum channel* is a linear mapping from one mixed state to another mixed state

$$\rho \to \mathcal{E}(\rho).$$

- *Quantum states* are represented by (column) vectors in the Hilbert space using Dirac's *ket vector* notation:

$$|\psi\rangle = \begin{pmatrix} \alpha \\ \beta \end{pmatrix}.$$

- $sgn(x)$ is the sign of the number x.

- The *tensor product* of two quantum states $|\psi\rangle = \sum_j \alpha_j |j\rangle$, $|\phi\rangle = \sum_k \beta_k |k\rangle$ is $|\psi\rangle \otimes |\phi\rangle = \sum_{j,k} = \alpha_j \beta_k (|j\rangle \otimes |k\rangle)$.

- The *trace* of a matrix A is the sum of its diagonal elements, $tr(A) = \sum_i A_{ii} = \sum_i \langle e_i | A | e_i \rangle$, where $|e_i\rangle$ is the basis vector with 1 at the ith index and 0 everywhere else.

- A complex square matrix U is *unitary* if its complex conjugate transpose $U^\dagger$ is also its inverse:

$$U^\dagger U = U U^\dagger = I.$$

- The system, or *wave function*, of a qubit can be written as a linear combination of basis states.

PART I

Building Blocks

CHAPTER 1

Introduction

Just 40 years ago, the connection between computer science and quantum mechanics was made. For the first time, scientists thought to build a device to realize information processing and computation using the extraordinary theory that governs the particles and nuclei that constitute our universe. Since then, we find ourselves time and again amazed by the potential computing power offered by quantum mechanics as we understand more and more about it. Some problems that are previously thought to be intractable now have efficient solutions with a quantum computer. This potential advantage stems from the unconventional approach that a quantum computer uses to store and process information. Unlike traditional digital computers that represent two states of a bit with the on and off states of a transistor switch, a quantum computer exploits its internal states through special quantum mechanical properties such as superposition and entanglement. For example, a quantum bit (qubit) lives in a combination of the 0 and 1 states at the same time. Astonishingly, these peculiar properties offer new perspectives to solving difficult computational tasks. This chapter is dedicated to a high-level overview of the rise of quantum computing and its disruptive impacts. More importantly, we highlight the computer scientists' role in the endeavor to take quantum computing to practice sooner.

1.1 THE BIRTH OF QUANTUM COMPUTING

Paul Benioff began research on the theoretical possibility of building a quantum computer in the 1970s, resulting in his 1980 paper on quantum Turing machines [2]. His work was influenced by the work of Charles Bennett on classical reversible Turing machines from 1973 [3].

In 1982, the Nobel-winning physicist Richard Feynman famously imagined building a quantum computer to tackle problems in quantum mechanics [4, 5]. The theory of quantum mechanics aims to simulate material and chemical processes by predicting the behavior of the elementary particles involved, such as the electrons and the nuclei. These simulations quickly become unfeasible on traditional digital computers, which simply could not model the staggering number of all possible arrangements of electrons in even a very small molecule. Feynman then turned the problem around and proposed a simple but bold idea: why don't we store information on individual particles that already follow the very rules of quantum mechanics that we try to simulate? He remarked:

> "*If you want to make a simulation of nature, you'd better make it quantum mechanical, and by golly it's a wonderful problem, because it doesn't look so easy.*"

The idea of quantum computation was made rigorous by pioneers including David Deutsch [6, 7] and David Albert [8]. Since then, the development of quantum computing has profoundly altered how physicists and chemists think about and use quantum mechanics. For instance, by inventing new ways of encoding a quantum many-body system as qubits on a quantum computer, we gain insights on the best quantum model for describing the electronic structure of the system. It gives rise to interdisciplinary fields like quantum computational chemistry. As recent experimental breakthroughs and theoretical milestones in quantum simulation are made, we can no longer talk about how to study a quantum system without bringing quantum computation to the table.

1.1.1   THE RISE OF A NEW COMPUTING PARADIGM

For computer scientists, the change that quantum computing brings has also been nothing short of astounding. It is so far the only new model of computing that is not bounded by the extended Church–Turing thesis [9, 10], which states that all computers can only be polynomially faster than a probabilistic Turing machine. Strikingly, a quantum computer can solve certain computational tasks drastically more efficiently than anything ever imagined in classical computational complexity theory.

It is not until the mid-1990s that the power of quantum computing was becoming fully appreciated. In 1993, Bernstein and Vazirani [9] demonstrated a quantum algorithm with exponential speedup over any classical algorithms, deterministic or randomized, for a computational problem named recursive Fourier sampling. Many more astonishing discoveries followed. In 1994, Dan Simon [10] showed another computational problem that a quantum computer has an exponential advantage over any classical computers.

Then in the same year, Peter Shor [11, 12] discovered that more problems, namely factoring large integers and solving discrete logarithms, also have efficient solutions on a quantum computer, far more so than any classical algorithms that are ever known. The implication of this discovery is breathtaking. Existing cryptographic codes encrypt today's private network communications, data storage, and financial transactions, relying on the fact that prime factorization for sufficiently large integers is so difficult that the most powerful digital supercomputers could take thousands or millions of years to compute. But the security of our private information could be under threat, should a quantum computer capable of running Shor's algorithm be built.

In 1996, another algorithm by Lov Grover was discovered [13]. Once again, a quantum algorithm is shown to provide improvement over classical algorithms, and in this case Grover's algorithm exhibits quadratic speedup for the problem of *unstructured database search* in which we are given a database and aim to find some marked items. For example, given an unordered set S of N elements, we want to find where $x \in S$ is located in the set. Classically, we need $O(N)$ accesses to the database in the worst case, while quantumly, we can do it with $O(\sqrt{N})$ accesses.

These are just a few examples of quantum algorithms that have been discovered. When implemented appropriately on a quantum computer, they offer efficient solutions to problems that seem to be intractable in the classical computing paradigm.

Building a quantum computer is, however, extremely challenging. When the idea was first proposed, no one knew how to build such powerful computers. To realize the computational power, we must learn to coherently control and manipulate highly-entangled, complex, physical systems to near perfection.

In the last 30 years or so, technologies for manufacturing quantum chips have significantly advanced. Today, we are at an exciting time where small- and intermediate-scale prototypes have been built. It marks a new era for quantum computing, as John Preskill, a long-time leader in quantum computing at Caltech, puts it, "we have entered the *Noisy Intermediate-Scale Quantum* (NISQ) era," [14] in which quantum computing hardware is becoming large and reliable enough to perform small useful computational tasks. Research labs from both academia and industry, domestic and abroad, are now eager to experimentally demonstrate a first application of quantum computers to some real-world problems that any classical computers would have a hard time solving efficiently.

1.1.2   WHAT IS A QUANTUM COMPUTER?

In a nutshell, a quantum computer is a computing device that stores information in objects called quantum bits (or qubits) and transforms them by exploiting certain very special properties from quantum mechanics. Despite the peculiarity in the behavior of quantum mechanical systems (e.g., particles at very small energy and distance scales), quantum mechanics is one of the most celebrated and well-tested theory for explaining those behaviors. Remarkably, the non-intuitive properties and transformations in quantum systems have significant computational significance, as they allow a quantum computer to operate on an exponentially large computational space.

In contrast, a traditional digital computer stores information in a sequence of bits, each of which takes two possible values, 0 or 1, represented by the on and off of a transistor switch, for example. To manipulate information, it sends an input sequence of bits in the form of electrical signals through integrated circuits (IC) to produce another sequence of bits. This process is deterministic and fast, thanks to the advanced technologies in IC fabrication. Computers today can afford billions of instructions per second, without worrying about experiencing an error for billions of device hours.

1.2   MODELS OF QUANTUM COMPUTATION

The approaches to quantum computing (QC) can be roughly split into three main categories: (i) analog QC, (ii) digital gate-based QC, and (iii) measurement-based QC.

1.2.1   ANALOG MODEL

In analog QC [15, 16], one gradually evolves the state of a quantum system using quantum operations that smoothly change the system such that the information encoded in the final system corresponds to the desired answer with high probability. When the quantum system is restricted to evolve slowly and remains in a ground state energy throughout the evolution, then this approach is typically referred to as "adiabatic quantum computing" [17, 18]. When the restriction is lifted and the system is allowed to interact with the thermal environment, it is referred to as "quantum annealing" [19, 20]. This analog approach is sought after by companies including D-Wave systems, Google, and others. However, whether or not existing quantum annealing devices achieve universal quantum computation or any quantum speedup remains unclear.

1.2.2   GATE-BASED MODEL

In digital QC, information is encoded onto a discrete and finite set of quantum bits (qubits), and quantum operations are broken down to a sequence of a few basic quantum logic gates. We obtain the correct answer with high probability from the digital measurement outcomes of the qubits. A digital QC is typically more sensitive to noise from the environment than an analog QC. For instance, qubit decoherence is usually considered undesirable in digital QC except sometimes during initialization and measurement, whereas in adiabatic QC, decoherence helps the system relax to the ground energy state [17, 21]. In the NISQ era, noise including qubit decoherence, imprecise control, and manufacturing defect has non-negligible detrimental effects and can accumulate when running long quantum algorithms, necessitating noise mitigation techniques to protect the information during the computation. These devices are called the "NISQ digital quantum computers." In principle, the discretization of information allows for the discretization of errors and use of redundancy to encode information, which give rise to the use of quantum error correction (QEC) to achieve system-level fault tolerance. However, the overheads of conventional QEC approaches are found to be inhibitory in the near term. Devices that implement QEC are called "fault-tolerant quantum computers." Throughout the remainder of the book, the discussion will be centered around NISQ digital QC. Nonetheless, the general principles and techniques introduced here are applicable to all types of quantum computers. We refer the interested readers to a number of pertinent textbooks, reviews, and theses [22–29].

1.2.3   MEASUREMENT-BASED MODEL

One example of measurement-based quantum computation (MBQC) is the cluster state model—see a short review in [30]. In this model of quantum computation, one initializes a number of qubits in the *cluster state*. The cluster state is represented by a graph, in which each node is a qubit initializes in $|+\rangle$ and each edge denotes a controlled-Z gate. The graph can have any topology, e.g., a 1-D chain, or a 2-D grid. The computation process involves measuring (in some measurement basis) some of the qubits in the cluster state. Some of the measurements are

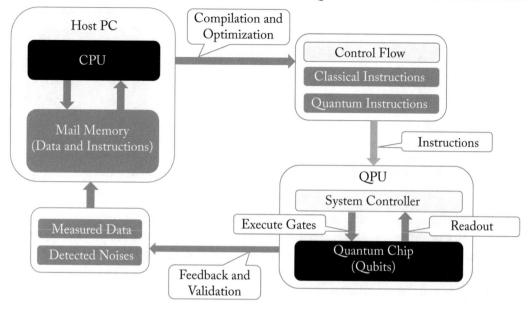

Figure 1.1: A QPU (quantum processor unit) and how it interacts with classical computers.

possibly conditioned on the outcomes of previous measured qubits. The key observation here is that each measurement equivalently accomplishes a quantum gate due to gate teleportation. The output of the computation is the measurement bit-string outcome and the remaining state of the qubits that are not measured. It is shown that this is a universal quantum computation model.

Our focus of the book is on the gate-based model; we present the other models of computation here for completeness, but details of the models are out of the scope of this book.

1.3 A QPU FOR CLASSICAL COMPUTING

Quantum computing hardware is currently envisioned to be a hardware accelerator for classical computers, as shown in Figure 1.1. To some extent, it is like a processing unit specialized in dealing with quantum information, in a way similar to a GPU (graphics processing unit) that specializes in numerical acceleration of kernels, including creation of images for display. For this reason, the QC hardware is referred to as a QPU (*quantum processing unit*). Unlike a GPU, which can perform arithmetic logic and data fetching at the same time, a QPU does not fetch data or instructions on its own. A host processor controls every move of the QPU. Let us now dive deeper into the architectural design of a QPU.

It is often misunderstood that a quantum computer is going to replace all classical digital computers. A quantum computer should *never* be viewed as a competitor with a classical com-

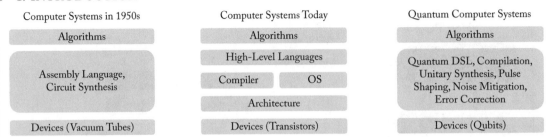

Figure 1.2: Architectural designs of classical vs. quantum computers. The abstraction layers for 1950s classical computing, today's classical computing, and quantum computing are compared.

puter. In fact, classical processing and classical control play vital roles in quantum computing. On one hand, a quantum algorithm generally involves classical pre- or post-processing. On the other hand, efficient classical controls are needed for running the algorithm on hardware. As such, a better way of regarding the QC hardware is as a co-processor or an accelerator, that is a QPU, as opposed to direct replacements of classical computers.

1.3.1   ARCHITECTURAL DESIGN OF A QPU

A quantum computer implements a fundamentally different model of computation than a modern classical computer does. It would be surprising if the exact design of a computer architecture would extend well for a quantum computer [31, 32]. As shown in Figure 1.2, the architectural design of a quantum computer resembles that of a classical computer in the 1950s where device constraints are so high that full-stack sharing of information is required from algorithms to devices. In time, as technology advances and resource becomes abundant, a quantum computer perhaps will adapt to the modularity and layering models as seen in classical architectures. But in the short term, as long as the NISQ era lasts, it is premature to copy the abstraction layers of today's conventional computer systems to a quantum system.

Furthermore, quantum information processing is fundamentally different than what computer engineers are used to. For instance, for conventional computers, engineers go to great lengths in minimizing the noises caused by quantum mechanics in the transistor components. Rather than suppressing its effects, a quantum computer harnesses the power of quantum mechanics. As such, the control apparatus for a quantum computer would look drastically different from that of a conventional computer.

In reality, for successful operation, a quantum computer must implement a well-isolated physical system that encodes a sufficiently large number of qubits, and controls these qubits with extremely high speed and precision in order to carry out computation. The rest of the section describes at a high level the key components in a fully functional quantum computer architecture. The developments of digital quantum computers, for both the NISQ and FT era, still face challenges, which comprise of reliably addressing and controlling qubits and correcting errors.

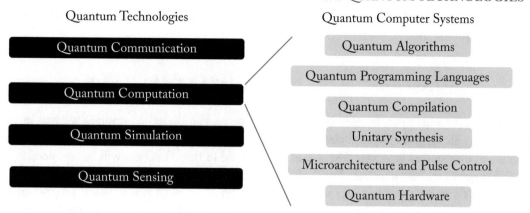

Figure 1.3: Quantum computation is one of the promising technologies made for harnessing the power of quantum systems.

The control complexity becomes overwhelming as the number of qubits scale up, necessitating system-level automation to guarantee the successful execution of quantum programs [33]. As such, classical computers are needed to control and assist the quantum processor. Quantum computers are generally viewed as co-processors or accelerators of classical computers, as shown in Figure 1.1.

To some extent, the quantum computer architecture illustrated above arguably resembles in-memory processing or reconfigurable computing architectures. As shown in Figure 1.1, inside a QPU, quantum data are implemented by physical (quantum mechanical) objects such as atoms while quantum gates are control signals such as lasers acting on the data—this "gates-go-to-data" model of computation motivates a control unit close to the quantum data and an interface that talks frequently with the quantum memory and the classical memory.

1.4   QUANTUM TECHNOLOGIES

The broad field of quantum technology encompasses more than just quantum computation; it can be roughly divided into four domains shown in Figure 1.3. (i) In *quantum computation*, quantum systems are carefully isolated and controlled to store and transform information in a way that is promised to be drastically more efficient than classical digital computers. (ii) *Quantum communication* [34–39] aims to use entangled photons to encrypt and transmit data securely. (iii) To study the structure and properties of electronic systems, *quantum simulation* [4, 40–46] maps the problem to a well-defined, controlled quantum system to mimic the behavior of quantum systems of interests. (iv) *Quantum sensing* [47–49] use quantum coherence to improve precision measurements of physical quantities. Each domain has its own focus, yet one can usually benefit from the techniques developed for another. Although mainly about quantum computation, this book will extend its discussions from time to time to the other domains of quantum technologies.

1.5   A ROAD MAP FOR QUANTUM COMPUTERS

Today's quantum computers resemble, in many aspects, the digital computers we had in the 1950s. They are large in physical size, limited in the number of computing units, expensive to build, and demanding in power. The machines we build in the NISQ era will be equipped with 50–1000 qubits and are capable of performing operations with error rate around 10^{-3} or 10^{-4}. The state-of-the-art quantum gate error rate is around 10^{-2}. As a result, when programming for a quantum computer, we have no choice but to optimize every bit of the limited resources. When qubits are not only short-lived but also limited in number and when quantum logic gates are noisy, every variable and every instruction in a quantum program matter.

In fact, we have been here before. After all, this is where we started for classical digital computers. The introduction of integrated circuits (IC) in the 1960s paved the ground work for the impressive performance growth of contemporary digital computers. In 1964, Gordon Moore accurately projected the exponential growth in the number of transistors per integrated circuit based on the cost of IC fabrication, now known as the Moore's Law. After half a century of investment and development in hardware, architecture, and software, we have built a computing ecosystem that has deeply changed our society and transformed the way we live, work and communicate [50].

Many believe that similar scaling will be achieved for quantum computers. For instance, the reported qubit coherence times (i.e., lifetime) for superconducting qubits have been on track for an encouraging exponential increase so far, following the so-called *Schoelkopf's Law*. But whether this scaling will last relies on continued investment in the field of quantum computing, driven by not only our scientific curiosity but also its economic and social impacts.

Fueled by joint efforts from research institutions and technology companies worldwide, progress in quantum hardware has been impressive. IBM [51, 52] and Google [53] are testing superconducting machines with more than 50 quantum bits (qubits) and providing users with cloud access to their prototypes. Intel [54, 55] is building quantum computers with silicon spin qubits and cryogenic controls. IonQ [56] has announced a 79-qubit "tape-like" trapped-ion quantum computer. Other multinational companies, including Intel, Microsoft, and Toshiba, are also making efforts toward practical-scale fully programmable quantum computers. Many others, although not building prototypes by themselves, are joining the force by investing in the field of quantum computing. Machines up to 100 qubits are around the corner, and even 1,000 qubits appears buildable. John Preskill notes that we are at a "privileged time in the history of science and technology" [14]. Specifically, classical supercomputers cannot simulate quantum machines larger than 50–100 quantum bits. Emerging physical machines will bring us into unexplored territory and will allow us to learn how real computations scale in practice.

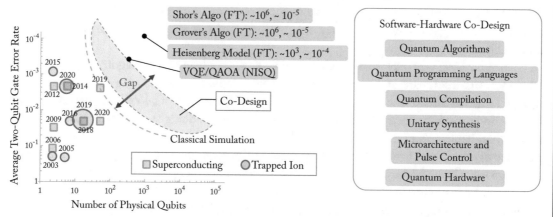

Figure 1.4: Status of qubit technologies [57–67]. Also drawn the gap between algorithms and realistic machines. Breaking abstractions via software-hardware co-design will be key in closing this gap for NISQ computers, hence the overarching theme of this book.

The key to quantum computation is that every additional qubit doubles the computational space in which the quantum machines operate. However, this extraordinary computing power is far from fully realized with today's technology, as the quantum machines will have high error rates for some time to come. Ideally in the long term, we would use the remarkable theory of quantum error correction codes to support error-free quantum computations. The idea of error correction is to use redundancy encoding, i.e., grouping many physical qubits to represent a single, fault-tolerant qubit. As a consequence, a 100 qubit machine can only support, for example, 3–5 usable logical qubits. Until qubit resources become much larger, another practical approach would be to explore error-tolerant algorithms and use lightweight error-mitigation techniques in the near term. So NISQ machines imply living with errors and exploring the effects of noise on the performance and correctness of quantum algorithms.

1.5.1   COMPUTER SCIENCE RESEARCH OPPORTUNITIES

Despite technology advances, there remains a wide gap between the machines we expect and the algorithms necessary to make full use of their power. In Figure 1.4, we can see the size of physical machines (in this case trapped ion machines) over time. Ground-breaking theoretical work produced Shor's algorithm [12] for the factorization of the product of two primes and Grover's algorithm [13] for quantum search, but both would require machines many orders of magnitude larger than currently practical. This gap has led to a recent focus on smaller-scale, heuristic quantum algorithms in areas such as quantum simulation, quantum chemistry, and quantum approximate optimization algorithms (QAOA) [68]. Even these smaller-scale algorithms, however, suffer from a gap of two to three orders of magnitude with respect to recent machines. Relying on solely technology improvements may require 10–20 years to close even this smaller gap.

A promising way to close this gap sooner is to create a bridge from algorithms to physical machines with a software-architecture stack that can increase qubit and gate efficiency through *automated optimizations and co-design* [31]. For example, recent work [69] on quantum circuit compilation tools has shown that automated optimization produces significantly more efficient results than hand-optimized circuits, even when only a handful of qubits are involved. The advantages of automated tools will be even greater as the scale and complexity of quantum programs grows. Other important targets for optimization are mapping and scheduling computations to physical qubit topologies and constraints, specializing reliability and error mitigation for each quantum application, and exploiting machine-specific functionality such as multi-qubit operators.

Quantum computing technologies have recently advanced to a point where quantum devices are large and reliable enough to execute some applications such as quantum simulation of small-size molecules. This is an exciting new era because being able to program and control small prototypes of quantum computers could lead to discoveries of more efficient algorithms to real-world problems than anything ever imagined in the classical computing paradigm. Public interests, including commercial and military interests, are essential in keeping substantial support for basic research. Recent discoveries of quantum applications in chemistry, finance, machine learning, and optimization are just some early evidence of a promising future ahead. Looking forward, we are on the track to continuously grow the performance of the quantum hardware, complemented with an efficient, scalable, and robust software toolflow that it demands [32, 70, 71].

CHAPTER 2

Think Quantumly About Computing

We begin this chapter with a presentation of the intuitions behind quantum information and quantum computation (Section 2.1). They then are made rigorous with mathematical formulations in Section 2.2.

2.1 BITS VS. QUBITS

In this section, the elements of classical computing are compared and contrasted with those of quantum computing (Section 2.1.1). The introduction to quantum mechanics in this section is tailored for computer scientists, assuming no prior knowledge in physics (Section 2.1.2). A number of architectural implications is then introduced (Section 2.1.3), arising from the special properties and transformations in this new computing paradigm.

2.1.1 COMPUTING WITH BITS: BOOLEAN CIRCUITS

Part of the learning curve of quantum computing (QC) stems from its unfamiliar nomenclature. Some is required for expressing the special properties of quantum mechanics, but the rest is merely a reformulation of what we already know about what an ordinary computer can do. As such, to prepare the reader for later discussion in QC, we briefly revisit how classical digital computers work, but in the language and notation used by QC. In particular, we will review four fundamental concepts in the classical theory of computing: the circuit model, von Neumann architecture, reversible computation, and randomized computation.

Boolean Circuits

A number of classical models of computation are developed to describe the components of a computer necessary to compute a mathematical function. Some familiar ones include the *Turing machine model* (sequential description), the *lambda calculus model* (functional description), etc. In this section, we choose to review the *Boolean circuit model* of computation, which is considered the easiest to extend to the theory of quantum computing. These models, although expressing computability and complexity from different perspectives, are in fact equivalent. Specifically, every function computable by an $n-$input Boolean circuit is also computable by a Turing ma-

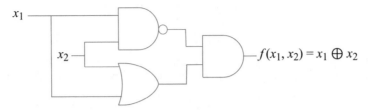

Figure 2.1: A Boolean circuit implementing the XOR function using a NAND gate, an OR gate, and an AND gate. Lines are wires that transmit signals, and shaped boxes are gates. Signals are copied/duplicated where wires split into two.

chine of length-n inputs, and vice versa. The *size* of a circuit, defined by the number of gates it uses, is closely related to the *running time* of a Turing machine.

In a classical digital computer (under the Boolean circuit model), information is stored and manipulated in bits—strings of zeros and ones, such as 10011101. The two states of each bit in the string are represented in the computer by a two-level system, such as charge (1) or no charge (0) in a memory cell (for storing) and high (1) or low (0) voltage signal in a circuit wire (for transmitting).

In the "bra-ket" notation invented by Paul Dirac in 1939, the state of a bit is denoted by the symbol, $|\rangle$. So, the two-level system can be written as $|0\rangle$ and $|1\rangle$, or $|\uparrow\rangle$ and $|\downarrow\rangle$, or $|charge\rangle$ and $|no\ charge\rangle$, etc. The above length-8 bit string can thus be written as $|1\rangle\,|0\rangle\,|0\rangle\,|1\rangle\,|1\rangle\,|1\rangle\,|0\rangle\,|1\rangle$, or $|10011101\rangle$ for short. Why is this called the "bra-ket" notation? In fact, $|\cdot\rangle$ is called the "ket" symbol and $\langle\cdot|$ is called the "bra" symbol, and together they form a bracket $\langle\cdot|\cdot\rangle$. Later, we will see in the linear algebra representation of quantum bits, they correspond to column vectors and row vectors, respectively. For now, the reader may regard this notation as pure symbolism—its advantages will be clear once we discuss operations of quantum bits.

Any computation can be realized as a circuit of *Boolean logic gates*. For example, the following is a circuit diagram for computing the XOR function of two input bits: $f(x_1, x_2) = x_1 \oplus x_2$ (Figure 2.1).

In this classical Boolean circuit, lines are "wires" that transmit signals, and boxes are "gates" that transform the signals. Signals are copied/duplicated at places where wires split into two. The above shows one possible implementation of the XOR function with AND, OR, and NAND gates. It is well known that the NAND gate, along with duplication of wires and use of ancilla bits (i.e., ancillary input bits typically initialized to 0), is universal for computation. In other words, any Boolean function is computable by "wiring together" a number of NAND gates.

The Boolean circuit model is a useful theoretical tool for analyzing what functions can be *efficiently* implemented. It is also a convenient tool for computer architects and electrical engineers as it is close to the physical realization of today's computers. The von Neumann architecture is one example of a design of modern computers.

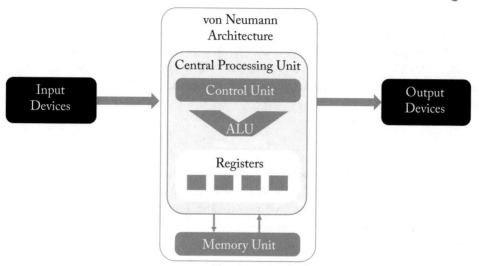

Figure 2.2: The von Neumann Architecture of a classical computer.

von Neumann Architecture

In the following we describe the key components comprising modern digital computers, first proposed by John von Neumann in 1945. In his description, a *von Neumann architecture* computer has these components, as shown in Figure 2.2: (i) a central processing unit (CPU) including an arithmetic logic unit (ALU) and a control unit; (ii) a random-access memory (RAM) that stores data and program instructions; (iii) input and output (I/O) devices; and (iv) external storage.

The instruction set architecture (ISA), serving as the interface between hardware and software, defines what a computer natively supports, including data types, registers, memory models, I/O support, etc. Modern ISAs are commonly classified into two categories: (i) complex instruction set computer (CISC) that supports many specialized operations regardless of how rarely they are used in a program. One example of CISC is the Intel x86-family architecture; and (ii) reduced instruction set computer (RISC) that includes only a small number of essential operations, such as the RISC-V architecture [72].

The CPU realizes (implements) the ISA. While its design can be very complex, the CPU typically has a control unit that fetches and executes instructions by directing signals accordingly, and an ALU that performs arithmetic and logic operations on data. Most modern CPUs are implemented in electric circuitry, as seen in the Boolean circuit model, printed on a flat piece of semiconductor material, known as an integrated circuit (IC).

Over the past few decades, production cost for IC has been drastically reduced thanks to advancement in technology [50]. We can build transistors, the building blocks of an IC, smaller and smaller, cheaper and cheaper. The number of transistors that can be economically printed per IC has been growing exponentially over time—approximately doubling every 1.5 years. This trend is referred to as the "Moore's Law." But as most believed, this trend is not a sustainable one, due to both physical limitations and market size. It is expected that within five years the feature size of transistors will stop at a few nanometers. As it approaches the atomic level (also on the order of nanometer), noises from quantum mechanical processes will start to dominate and perturb the system.

Reversible Computation

The study of reversible computing originally arises from the motivation to improve the computational energy efficiency in the 1960s and 1970s led by Laundauer [73] and Bennet [74]. Quantum computers transform quantum bits reversibly (except for initialization and measurement). The connection between reversible computation and quantum mechanics was discovered by Benioff in the 1980s [2]. As a result, QC benefits a great deal from the study of reversible computing, and vice versa. Later, we will see the roles of reversible computing in quantum circuits.

According to the second law of thermodynamics, an irreversible bit operation, such as the OR gate, must dissipate energy, typically in the form of heat. Specifically, suppose the output of an OR gate is $|1\rangle$. We cannot infer what the inputs were—they could be anything from $|01\rangle$, $|10\rangle$, or $|11\rangle$. The von Neumann–Landauer limit states that $kT\ln(2)$ energy is dissipated per irreversible bit operation. However, some bit operations are theoretically (logically) "reversible"— in the sense of that the output state uniquely determines the input state of the operation. For example, the NOT gate is reversible. Flipping the state of a bit from $|0\rangle$ to $|1\rangle$, or vice versa, does not create or erase information from the system. To some extent, reversible also means time-reversible—the transformation done by a reversible circuit can be reverted by applying the inverse transformation (which always exists).

One could imagine a computer can be built consisting solely of reversible operations. In analogy to the NAND gate being universal for Boolean logic, is there a reversible gate set that is universal? The answer is yes. To illustrate this, we introduce three example reversible gates: the NOT gate, the CNOT (controlled-not) gate, and the Toffoli (controlled-controlled-not) gate, all of which are self-inverse (i.e., applying the gate twice returns the bits to their original state). Their Boolean circuit notations and truth-tables can be found in Table 2.1. Specifically, the NOT gate negates the state of the input bits. The CNOT gate is a conditional gate—the state of the target bit x_2 is flipped if the control bit x_1 is $|1\rangle$. It is the reversible version of the XOR gate. The Toffoli gate has two control bits, x_1 and x_2, and one target bit x_3. Similarly, the target bit is flipped if both the control bits are $|1\rangle$. It is particularly handy as it can be used to simulate the NAND gate and the DUPE gate (with the use of ancillas), and thus is a universal reversible gate.

More formally, we note that the Toffoli gate is universal, in that any (possibly non-reversible) Boolean logic can be simulated with a circuit consisting solely of Toffoli gates, given

Table 2.1: Reversible logic gates. The truth table of reversible logic gates shows the permutation of bits. Toffoli is universal reversible computation.

Reversible Gate	Boolean Circuit Notation	Truth Table
NOT gate	x_1 —[X]— $\mathrm{NOT}(x_1)$	$\lvert 0\rangle \mapsto \lvert 1\rangle$ $\lvert 1\rangle \mapsto \lvert 0\rangle$
CNOT gate	x_1 —•— x_1 x_2 —⊕— $x_1 \oplus x_2$	$\lvert 00\rangle \mapsto \lvert 00\rangle$ $\lvert 01\rangle \mapsto \lvert 01\rangle$ $\lvert 10\rangle \mapsto \lvert 11\rangle$ $\lvert 11\rangle \mapsto \lvert 10\rangle$
Toffoli gate	x_1 —•— x_1 x_2 —•— x_2 x_3 —⊕— $\mathrm{AND}\,(x_1, x_2) \oplus x_3$	$\lvert 000\rangle \mapsto \lvert 000\rangle$ $\lvert 001\rangle \mapsto \lvert 001\rangle$ $\lvert 010\rangle \mapsto \lvert 010\rangle$ $\lvert 011\rangle \mapsto \lvert 011\rangle$ $\lvert 100\rangle \mapsto \lvert 100\rangle$ $\lvert 101\rangle \mapsto \lvert 101\rangle$ $\lvert 110\rangle \mapsto \lvert 111\rangle$ $\lvert 111\rangle \mapsto \lvert 110\rangle$

that ancilla inputs and garbage outputs are allowed. Proof of this theorem is omitted here. As such, a generic reversible circuit has the form shown in Figure 2.3.

In this circuit, a Boolean function $f : \{0, 1\}^n \mapsto \{0, 1\}^m$ is computed reversibly using only Toffoli gates. All ancilla inputs are initialized to $\lvert 1\rangle$ (if needed, $\lvert 0\rangle$ ancilla can be produced as well, because a Toffoli gate on $\lvert 111\rangle$ gives $\lvert 110\rangle$). All garbage bits will be discarded at the end of the circuit.

One cannot overemphasize the above theorem's implication to quantum computing—as noted before, a quantum computer transforms quantum bits reversibly, so this theorem implies that any Boolean circuit can be transformed into a reversible one, and then a quantum one by implementing a quantum Toffoli gate and replacing each bit with a quantum bit. Reversible circuit synthesis is a useful tool in designing quantum circuits.

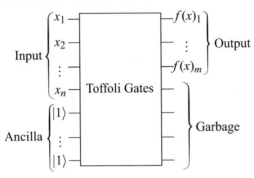

Figure 2.3: A generic reversible circuit for implementing a possibly irreversible function f : $\{0, 1\}^n \rightarrow \{0, 1\}^m$.

Randomized Computation

So far, we have not discussed one familiar ingredient to the computation that appears commonly in classical computing—randomness. Many natural processes exhibit unpredictable behavior, and we should be able to take advantage of this unpredictability in computation and algorithms. The notion of probabilistic computation seems realistic and necessary. On one hand, the physical world contains randomness, as commonly seen in quantum mechanics. On the other hand, we can propose several computational problems that we do not yet know how to solve efficiently without randomness. If BPP=P, however (i.e., the complexity class *bounded-error probabilistic polynomial time* is equivalent to the class *deterministic polynomial time*), as some believe, then randomness is unnecessary and we can simulate randomized algorithms as efficiently with deterministic ones. Nonetheless, randomness is still an essential tool in modeling and analyzing the physical world. We can find many examples where randomness is useful: in economics, it is well known that Nash equilibrium always exists if players can have probabilistic strategies, and in cryptography, a secret key typically relies on the uncertainty in itself.

Randomness as a resource is typically used in computation in the following two forms: (i) an algorithm can take random inputs; and (ii) an algorithm is allowed to make random choices. As such, we introduce the notion of random bits and coin flips, again in the "bra-ket" and circuit notations.

Suppose x_1 is a random bit, and the state of x_1 is $|0\rangle$ with probability $\frac{1}{2}$ and $|1\rangle$ with probability $\frac{1}{2}$, denoted as:

$$|x_1\rangle = \frac{1}{2}|0\rangle + \frac{1}{2}|1\rangle.$$

For now, this notation may look strange and cumbersome, but the benefit of writing the state of a bit this way will become clear when we generalize to the quantum setting. The state is called a probability distribution of $|0\rangle$ and $|1\rangle$. To describe a general n-bit probabilistic system, we write down the underlying state of the system as:

$$\sum_{b \in \{0,1\}^n} p_b \, |b\rangle \,,$$

where b is any possible length-n bit-string, and p_b is called the probability of b. By basic principles of probability, all p_b values must be non-negative and summing to 1.

In reality, the physical system is in one of those possible state. When we execute a randomized algorithm, we expect to *observe* (sample) the outcome at the end. From the observer's perspective, the values of the random bits are uncertain (hidden) until they are observed. Once some of the random bits in the system are observed, then the state of the system (to the observer's knowledge) is changed to reflect what was just learned, following laws of *conditional probability*. For example, a random system can be described with:

$$|x_1 x_2\rangle = \frac{1}{8} |00\rangle + \frac{1}{4} |01\rangle + \frac{5}{8} |10\rangle + 0 |11\rangle \,.$$

Now suppose it is observed that the first bit is $|0\rangle$ (the probability of this scenario is $\mathbf{Pr}[x_1 = 0] = \frac{1}{8} + \frac{1}{4} = \frac{3}{8}$). The state of the system *after* the observation is then conditioned on our observation:

$$|x_1 x_2\rangle \text{ (given } x_1 = 0) = \frac{1/8}{3/8} |00\rangle + \frac{1/4}{3/8} |01\rangle = \frac{1}{3} |00\rangle + \frac{2}{3} |01\rangle \,.$$

Here the bit-strings inconsistent with the outcome are eliminated, and the remaining ones are renormalized.

In a randomized algorithm, we typically allow that (i) it is correct with high probability, or (ii) it does not always run in desired time. Some of the uncertainty comes from its ability to make decisions based on the outcome of a coin flip. Now suppose we have implemented a conditional-coin-flip gate, named CCOIN: —[CCOIN]—. When the input bit is $|0\rangle$, CCOIN does nothing. When the input is $|1\rangle$, CCOIN tosses a fair coin:

$$\text{CCOIN} = \begin{cases} |0\rangle \mapsto |0\rangle \,, \\ |1\rangle \mapsto \frac{1}{2} |0\rangle + \frac{1}{2} |1\rangle \,. \end{cases}$$

Suppose we have a random program that reads: (1) Initialize a bit to $x_1 = |1\rangle$. (2) Flip a fair coin if x_1 is $|1\rangle$ and write result to x_1. (3) Repeat step 2. In terms of a circuit, the program looks like:

$$x_1 : |1\rangle\!-\!\boxed{\text{CCOIN}}\!-\!\boxed{\text{CCOIN}}\!-\!$$

One is interested in observing the outcome at the end of the program. Let's analyze the circuit step by step. After the first CCOIN gate, $|x_1\rangle$ is set to $|0\rangle$ and $|1\rangle$ with equal probability (i.e., $|x_1\rangle = \frac{1}{2} |0\rangle + \frac{1}{2} |1\rangle$). After the second CCOIN gate, the state becomes $|x_1\rangle = \frac{1}{2} |0\rangle + \frac{1}{2}(\frac{1}{2} |0\rangle + \frac{1}{2} |1\rangle) = \frac{3}{4} |0\rangle + \frac{1}{4} |1\rangle$. It is convenient to write the above process in a state transition diagram:

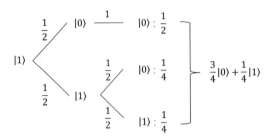

The following is a slightly larger circuit, where a CNOT gate correlates the two bits.

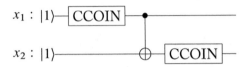

The system is initialized in $|11\rangle$. After the first CCOIN gate, the system is put into a random state: $\frac{1}{2}|01\rangle + \frac{1}{2}|11\rangle$. The CNOT gate then transforms the system to $\frac{1}{2}|01\rangle + \frac{1}{2}|10\rangle$, correlating the two bits. And finally after the second CCOIN gate: $|x_1 x_2\rangle = \frac{1}{2}(\frac{1}{2}|00\rangle + \frac{1}{2}|01\rangle) + \frac{1}{2}|10\rangle = \frac{1}{4}|00\rangle + \frac{1}{4}|01\rangle + \frac{1}{2}|10\rangle$. Again with a state transition diagram:

2.1.2 COMPUTING WITH QUBITS: QUANTUM CIRCUITS

Finally, we present to the reader, as efficiently as possible, the fundamental concepts in the quantum mechanics model of computation. Many believe that quantum computing can be described simply as randomized computing with a twist where we allow the "probability" to take negative (possibly complex) values. Alternatively, it can also be described as reversible computing with an additional "Hadamard" gate. Therefore, the goal of this section is to argue the meanings and implications of these statements.

Quantum Circuit Model

As usual, we start by describing the state of the quantum system ψ using the "bra-ket" notation introduced earlier. Suppose ψ is an n-qubit (quantum bit) system:

$$|\psi\rangle = \sum_{b\in\{0,1\}^n} \alpha_b |b\rangle,$$

where the coefficient α_b is called the *amplitude* (as opposed to probability) of the basis bit-string b. Just like probabilities, the amplitudes have two constraints: (i) they can take any complex numbers; and (ii) their sum of squared values is 1: $\sum_{b\in\{0,1\}^n} |\alpha_b|^2 = 1$. In the context of qubits, the probability distribution across bit-strings is called the *superposition* of all bit-strings; the correlation between bits is called *entanglement* of qubits. It is important to note that these are *not* renamings of the same concepts[1]—as random bits and quantum bits are fundamentally different objects. Despite the striking parallelism between the two, we should always be wary of the subtleties that differentiate them when analyzing a random circuit vs. a quantum circuit.

To *measure* (observe) the outcome of a qubit, we follow almost exactly what we did with a random bit. For an n-qubit system, if we measure all qubits at the end of a circuit,[2] then from $|\psi\rangle = \sum_{b\in\{0,1\}^n} \alpha_b |b\rangle$, we observe the bit-string $|b\rangle$ with probability $|\alpha_b|^2$. Upon measurement, the state of the system "collapses" to the single classical definite value: $\text{Meas}(|\psi\rangle) = |b\rangle$, and can no longer revert to the superposition as it was before. For example, the superposition state $|\psi\rangle = \frac{1}{\sqrt{2}}|0\rangle + \frac{1}{\sqrt{2}}|1\rangle$ yields, upon measurement, either outcome with equal probability: $\mathbf{Pr}[\text{Meas}(|\psi\rangle) = |0\rangle] = \mathbf{Pr}[\text{Meas}(|\psi\rangle) = |1\rangle] = \frac{1}{2}$.

One operation that is of fundamental importance to quantum computation is called the *Hadamard transformation*, a single-qubit quantum gate denoted as —$\boxed{\text{H}}$— in the circuit model:

$$\text{H} = \begin{cases} |0\rangle \mapsto \frac{1}{\sqrt{2}}|0\rangle + \frac{1}{\sqrt{2}}|1\rangle, \\ |1\rangle \mapsto \frac{1}{\sqrt{2}}|0\rangle - \frac{1}{\sqrt{2}}|1\rangle. \end{cases}$$

It turns out that allowing Hadamard gates in a reversible circuit (consisting of Toffoli gates) extends the circuit model over to any functions allowed to be computed on qubits (up to global phase). For this reason, H gate together with Toffoli gate are universal for quantum computation. Note that it does not mean that Nature allows only Hadamard and Toffoli transformations on qubits—as we will see in later sections, the laws of quantum mechanics allow a class of transformations, called *unitary transformations*.

One would argue that any interesting quantum mechanics phenomenon can be explained by *interference*. Unlike probability values that are always non-negative, amplitudes (as being possibly negative) can either accumulate and cancel. When two amplitudes accumulate, we say they

---

[1]Many believed that quantum mechanics has deterministic explanations, notably by the argument from EPR paradox (by Einstein, Podolsky, and Rosen in 1935 [75]) and other hidden-variable theories which try to equalize statistical correlation with entanglement. But later in 1964, John Bell famously showed Bell's theorem [76] that disproved the existence of local hidden variables of some types.

[2]This is a reasonable assumption by the law of deferred measurement.

interfere constructively; when they cancel each other out, we say they interfere destructively. The example circuit below illustrates this phenomenon:

$$x_1 : |0\rangle - \boxed{H} - \boxed{H} -$$

As usual, let's analyze the circuit step by step. After the first H gate, $|x_1\rangle$ is set to a superposition state $|x_1\rangle = \frac{1}{\sqrt{2}}|0\rangle + \frac{1}{\sqrt{2}}|1\rangle)$. After the second H gate, the state becomes $|x_1\rangle = \frac{1}{\sqrt{2}}(\frac{1}{\sqrt{2}}|0\rangle + \frac{1}{\sqrt{2}}|1\rangle) + \frac{1}{\sqrt{2}}(\frac{1}{\sqrt{2}}|0\rangle - \frac{1}{\sqrt{2}}|1\rangle) = |0\rangle$. Note that in this circuit, amplitudes of $|1\rangle$ cancel each other out (destructively interfere), while those of $|0\rangle$ accumulate (constructively interfere).

Again we track the state of the qubits as the circuit runs (from left to right), using a transition diagram. In the context of qubit states, the diagram is called the *Feynman Path diagram*, named after physicist Richard Feynman:

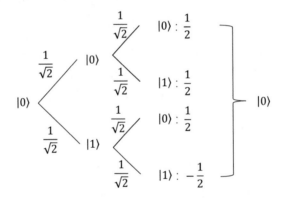

In this diagram, the state of the qubits also evolves from left to right.

Feynman's Sum-Over-Path Approach

Here we describe the precise prescription to track the amplitudes of a quantum state using Feynman's "sum-over-path" approach. The idea comes from the well-known theory of *path integral* [77, 78]. His key observation is that the final amplitudes of a quantum state can be written as a weighted sum over all possible paths the quantum system can take from the initial to the final state. In particular:

- the final amplitude is given by adding the contributions from all paths; and

- the contribution from a path is given by multiplying the coefficients along the path.

In the above example, there are a total of four paths (from left to right):

1. $|0\rangle \to |0\rangle \to |0\rangle$: $\frac{1}{\sqrt{2}} \times \frac{1}{\sqrt{2}} = \frac{1}{2}$

2. $|0\rangle \to |0\rangle \to |1\rangle$: $\frac{1}{\sqrt{2}} \times \frac{1}{\sqrt{2}} = \frac{1}{2}$

3. $|0\rangle \to |1\rangle \to |0\rangle$: $\frac{1}{\sqrt{2}} \times \frac{1}{\sqrt{2}} = \frac{1}{2}$

4. $|0\rangle \to |1\rangle \to |1\rangle$: $\frac{1}{\sqrt{2}} \times \frac{-1}{\sqrt{2}} = -\frac{1}{2}$

The amplitude of $|0\rangle$ is obtained from adding path 1 and path 3, while that of $|1\rangle$ is obtained from adding path 2 and path 4.

One neat trick to *prevent* interference is by introducing an entangled ancilla qubit, such as the following:

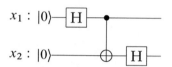

In particular, the qubits are initialized to all zero state. After the first H gate and CNOT gate, we arrive at $|x_1 x_2\rangle = \frac{1}{\sqrt{2}}|00\rangle + \frac{1}{\sqrt{2}}|11\rangle$. In fact, this state is an example of a "Bell state" due to John Bell [76] (or an "EPR pair" named after Einstein, Podolsky, and Rosen [75]), a class of entangled states. With the Bell state, we now apply the second H gate. At the end of the circuit, we obtain $|x_1 x_2\rangle = \frac{1}{\sqrt{2}}(\frac{1}{\sqrt{2}}|00\rangle + \frac{1}{\sqrt{2}}|01\rangle) + \frac{1}{\sqrt{2}}(\frac{1}{\sqrt{2}}|10\rangle - \frac{1}{\sqrt{2}}|11\rangle)$. This process is again illustrated with the following Feynman path diagram:

$|00\rangle$

$\frac{1}{\sqrt{2}}$ $|01\rangle$ —— $|00\rangle$

$\frac{1}{\sqrt{2}}$ $|10\rangle$ —— $|11\rangle$

$\frac{1}{\sqrt{2}}$ $|00\rangle$: $\frac{1}{2}$

$\frac{1}{\sqrt{2}}$ $|01\rangle$: $\frac{1}{2}$

$\frac{1}{\sqrt{2}}$ $|10\rangle$: $\frac{1}{2}$

$\frac{1}{\sqrt{2}}$ $|11\rangle$: $-\frac{1}{2}$

$\frac{1}{2}|00\rangle + \frac{1}{2}|01\rangle + \frac{1}{2}|10\rangle - \frac{1}{2}|11\rangle$

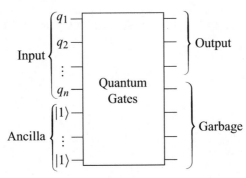

Figure 2.4: A generic quantum circuit that implements a unitary transformation, mapping from a quantum state storing the input (and ancilla) to a quantum state storing the output (and garbage).

Quantum Entanglement between Two Qubits

The Bell state, $|x_1 x_2\rangle = \frac{1}{\sqrt{2}}|00\rangle + \frac{1}{\sqrt{2}}|11\rangle$, is of particular interest: if we pick any of the two qubits to measure, we would obtain outcome $|0\rangle$ or $|1\rangle$ with equal probability; and the other qubit is guaranteed to be measured in the same state as the first. This "correlation" between the two qubits are called *quantum entanglement*. However, this is not to be confused with two correlated random bits each with equal probability of being observed $|0\rangle$ or $|1\rangle$. Proof of the distinction is omitted for the sake of brevity; we refer the interested reader to the studies on "local hidden variables" theory, Bell's inequality [76, 79], and the CHSH game [80]. In essence, the measurement of one qubit intrinsically alters and determines the state of the other. Perhaps even more surprising is that these two qubits can be physically far apart as long as they were previous entangled as a Bell state. To form such a relationship between the two qubits, they must interact with each other, either directly through a two-qubit gate as shown above, or indirectly through a photon or a third qubit. We will explain in mathematical terms the meaning of entanglement in the next section. Also see [81] for an extensive review.

A typical quantum circuit will look very much like a reversible circuit, except that the qubits are acted on by quantum gates (i.e., unitary transformations) as shown in Figure 2.4.

To illustrate how quantum circuits are implemented in hardware, we encourage the reader to avoid thinking about lines in a circuit as wires that carry electric signals about the bits, but rather, as qubit registers (i.e., physical objects) that store data over time. In particular, in the quantum circuit notation, qubits are denoted as lines, and gates are denoted as boxes, which are applied to qubits in the order from left to right. Physically, a qubit is a physical object integrated on the quantum chip (such as an atom, or a superconducting gadget) and a gate is pulse signals addressed to the qubits (such as a laser beam, or a microwave pulse). This is to be contrasted with a classical architecture, where a gate is a electric circuit component on the CPU and a bit is a voltage signal along wires sent through gates and switches.

2.1.3   ARCHITECTURAL CONSTRAINTS OF A QUANTUM COMPUTER

So far, we have discussed a number of exciting properties of qubits that together give rise to the power of a quantum computer, including superposition, entanglement, interference, etc. However, these properties are not easy to achieve—they incur strict requirements on the design of a quantum computer architecture, that is, the software and hardware structures of a quantum computer in practice.

1. *Probabilistic outcomes.* Information in a quantum state is extracted through statistics from measurements. Measuring an n-qubit system at the end of a program will not yield the full information about the quantum state, instead it reads out one of the 2^n possible bit-strings randomly. For this reason, a quantum program is intrinsically probabilistic. A good quantum program will make sure the desired bit-strings are observed with much higher probability compared to the undesired ones. Since it is not easy to entirely eliminate the probability of obtaining an undesired result, a program must be executed multiple times so as to gain statistical confidence on the observed outcome. Depending on the algorithm, a quantum program may need thousands of shots before meaningful statistics about the quantum state can be drawn. Algorithm designers typically use generic subroutines such as *amplitude amplification*, which improves the distribution by increasing the probability of the desired outcomes. The process for learning the full state information is called *quantum tomography*. Estimating the full quantum state of n qubits requires $O(2^{2n})$ shots of measurement.

2. *No copying of qubits.* Another fundamental limitation of a quantum computer system is the inability to make identical copies of an unknown qubit. This is called the *no-cloning theorem* due to Wootters and Zurek [82]. Duplication of bits, as introduced in classical Boolean circuit model, is prohibited in a quantum circuit. In classical computing, we are used to making copies of data when designing and programming algorithms. But in a quantum system, we can no longer easily read from or write to the quantum memory, as read is done through measurements (which likely alter the data) and write generally requires complex state preparation routines. The no-cloning limitation also prevents us from directly implementing a quantum analog of the classical memory hierarchy, as caches require making copies of data. Hence, current quantum computer architecture proposals follow the general principles that transformations are applied directly to quantum memory, and data in memory are moved but not copied. Two similar scenarios not to be confused with the no-cloning theorem are: (i) we are allowed to make an *entangled* copy of a qubit. In fact, for any arbitrary unknown state $|\psi\rangle = \alpha |0\rangle + \beta |1\rangle$, $\text{CNOT}(|\psi\rangle, |0\rangle)$ results in the entangled state $\alpha |00\rangle + \beta |11\rangle$. Measuring any of the two qubits would yield the same

statistical distribution; and (ii) we are allowed to rename (make alias of) a qubit when writing a quantum program.

3. *Qubit-qubit interactions.* Entanglement is arguably one of the essential concepts that provide a quantum computer its computational power. Without entanglement, an n-qubit quantum system is no better than $2n$ random variables (as every qubit can be modeled independently by two amplitudes). One could imagine any interesting quantum algorithm that explores exponential computational space by entangling its qubits in the system. In general, qubit-qubit interactions involve (i) applying directly complex multi-qubit gates, or (ii) if the coupling cannot be done, then either moving the qubits, or interacting through intermediate photons or qubits. This *communication cost* contributes to the very high implementation cost of quantum algorithms on NISQ machines. On one hand, complex multi-qubit gates are the dominant source of error as they are hard to achieve. On the other hand, today's NISQ machines typically have poor connectivity (i.e., only few qubits are allowed to interact directly). The design of a NISQ architecture is therefore inevitably focused on at reducing the overhead of qubit-qubit interactions.

4. *Analog noises.* A programmer seldom needs to concern failures in memory cells caused by external noises or gate operation mistakes in a conventional processor. This is because classical computer systems today are robust enough against environmental noises. Take the modern Intel Xeon Phi processor as an example. A 2017 data [83] shows that the "soft-error-rate" (charge disturbance from radiation that causes to flip the data state of memory cell, register, batch, or flip-flop) is around 100 FIT ("failure-in-time"), i.e., 100 errors per billion device hours (114,077 years). The experiments were performed in 500 hours running HPC (High Performance Computing) applications under a neutron beam (roughly 57,000 years of equivalent natural exposure). On the other hand, in stark contrast, NISQ machines are highly sensitive to environment and control noise—a 2020 data on the IBM Q Melbourne device (with 14 qubits) [84] shows that the average single-qubit gate error rate is 4.78×10^{-3}, the average two-qubit gate error rate is 9.46×10^{-2}, and the readout error rate is 8.03×10^{-2}. Furthermore, a qubit decoheres (loss of quantum information) naturally—the average T1 (T2) decoherence time for the above machine is about 50 μs (66 μs), due to spontaneous loss of energy (loss of phase). Each of these errors by itself may be have small effects on the quantum state, but if not mitigated or corrected, they can accumulate and become detrimental as we run long quantum programs. The effects of noise are so significant that a NISQ computer architecture must find strategies to reduce them.

2.2   BASIC PRINCIPLES OF QUANTUM COMPUTATION

Now we present a more rigorous picture of the central concepts in quantum computation. The theory of quantum computation can be formulated as a neat branch of mathematics. If we consider classical computation as operations under the laws of *boolean algebra*, then quantum computation operates under the rules of *linear algebra*. The probabilistic nature of quantum states adds another layer of complexity to understanding the behavior of quantum computers. Nonetheless, all of it can be beautifully captured in four simple postulates, describing quantum states, composition of quantum systems, measurements, and quantum gates, respectively. This mathematical formulation allows us to reason about how a quantum system behaves under our manipulation, i.e., *quantum logic* in a quantum computer. Throughout this section, basic linear algebra and probability theory concepts are reviewed or referenced when necessary.

Together, the four postulates describe how information is stored and manipulated in a quantum system. In particular, a quantum computer works with a finite set of computational objects called quantum bits (or *qubits*). The *quantum state postulate* defines the superposition state of each qubit. The *composition postulate* then generalizes it to represent a system of multiple qubits, and provides a formal, mathematical definition of the entanglement property. The *measurement postulate* is used to describe how much information can be read out from a quantum system, as well as the consequence of the measurement action to the system. Finally, the *quantum gate postulate* defines the logical operations that transform a quantum system.

2.2.1   QUANTUM STATES

Definition 2.1 (Superposition).   A single-qubit quantum state $|\psi\rangle$ can be defined as a (column) vector of two complex numbers:

$$|\psi\rangle = \begin{pmatrix} \alpha \\ \beta \end{pmatrix},$$

where $\alpha, \beta \in \mathbb{C}$ and $|\alpha|^2 + |\beta|^2 = 1$. Here, α and β are called the *amplitudes* of the quantum state. It is called a superposition state because we can rewrite it as a *linear combination* of the basis states $|0\rangle = \begin{pmatrix} 1 \\ 0 \end{pmatrix}$ and $|1\rangle = \begin{pmatrix} 0 \\ 1 \end{pmatrix}$ as follows:

$$|\psi\rangle = \alpha |0\rangle + \beta |1\rangle = \alpha \begin{pmatrix} 1 \\ 0 \end{pmatrix} + \beta \begin{pmatrix} 0 \\ 1 \end{pmatrix} = \begin{pmatrix} \alpha \\ \beta \end{pmatrix}.$$

Example 2.2   The two most common states are probably $|0\rangle$ and $|1\rangle$, often referred to as the *computational basis states*. Here, we highlight a few more quantum states that appear fairly frequently in quantum algorithms. For example, the "plus" and "minus" states:

$$|+\rangle = \frac{1}{\sqrt{2}}(|0\rangle + |1\rangle), \quad |-\rangle = \frac{1}{\sqrt{2}}(|0\rangle - |1\rangle).$$

And two other states with complex amplitudes:

$$|i\rangle = \frac{1}{\sqrt{2}}(|0\rangle + i\,|1\rangle), \quad |-i\rangle = \frac{1}{\sqrt{2}}(|0\rangle - i\,|1\rangle).$$

Last, we write the conjugate transpose of $|\psi\rangle$ as a row vector

$$\langle\psi| = (|\psi\rangle)^\dagger = \begin{pmatrix} \alpha^* & \beta^* \end{pmatrix}.$$

We can check the condition on the amplitudes by the inner product

$$\langle\psi|\psi\rangle = \alpha\alpha^* + \beta\beta^* = |\alpha|^2 + |\beta|^2 = 1.$$

More generally, we can extend to a *qudit system*: a d-dimensional qudit system is defined as a superposition of d basis states:

$$|\psi\rangle = \alpha_0\,|0\rangle + \alpha_1\,|1\rangle + \cdots + \alpha_{d-1}\,|d-1\rangle,$$

where $|\alpha_0|^2 + \cdots + |\alpha_{d-1}|^2 = 1$. In theory, we can construct a qudit system using qubits. However, in practice, many quantum systems are intrinsically a multi-level system. For example, a superconducting transmon has infinite levels among which the first few levels are easily accessible. A three-dimensional qudit system is sometimes called a *qutrit* system.

2.2.2   COMPOSITION OF QUANTUM SYSTEMS

Now, we illustrate how to represent a system consisting of multiple qubits. In classical computing, when moving from a single-bit system to a system consisting of n number of bit, we use a string of bits to represent the 2^n possible states that the system could be in, for there are exactly 2 choices for each bit. Take two bits, there are four possible states, namely 00, 01, 10, and 11. Naturally, intuition from the superposition principle tells us that, in a quantum computer, the *joint state* of a two-qubit system should be a linear combination of the *four* possible basis states, i.e., $|\psi\rangle = \alpha\,|00\rangle + \beta\,|01\rangle + \gamma\,|10\rangle + \delta\,|11\rangle$. Indeed, we can build a bigger quantum state from small quantum states using a *tensor product*.

Definition 2.3   (Composition).   The joint state of two separate quantum systems $|\psi_0\rangle = \sum_j \alpha_j\,|a_j\rangle$ and $|\psi_1\rangle = \sum_k \beta_k\,|b_k\rangle$ is represented as the *tensor product* of the components. That is,

$$|\psi\rangle = |\psi_0\rangle \otimes |\psi_1\rangle = \sum_j \sum_k \alpha_j \beta_k(|a_j\rangle \otimes |b_j\rangle),$$

where $|a_j\rangle \otimes |b_j\rangle$ can often be shortened as $|a_j b_k\rangle$.

Example 2.4   The four basis states of the two-qubit system are

$$|00\rangle = |0\rangle \otimes |0\rangle = \begin{pmatrix} 1 \\ 0 \\ 0 \\ 0 \end{pmatrix}, |01\rangle = |0\rangle \otimes |1\rangle = \begin{pmatrix} 0 \\ 1 \\ 0 \\ 0 \end{pmatrix},$$

$$|10\rangle = |1\rangle \otimes |0\rangle = \begin{pmatrix} 0 \\ 0 \\ 1 \\ 0 \end{pmatrix}, |11\rangle = |1\rangle \otimes |1\rangle = \begin{pmatrix} 0 \\ 0 \\ 0 \\ 1 \end{pmatrix}.$$

Example 2.5   Let's take a look at the example of two generic qubits. Suppose the first qubit is $|\psi_0\rangle = \alpha_0 |0\rangle + \alpha_1 |1\rangle$ and the second qubit is $|\psi_1\rangle = \beta_0 |0\rangle + \beta_1 |1\rangle$, then their joint state is:

$$|\psi\rangle = |\psi_0\rangle \otimes |\psi_1\rangle = \begin{pmatrix} \alpha_0 \\ \alpha_1 \end{pmatrix} \otimes \begin{pmatrix} \beta_0 \\ \beta_1 \end{pmatrix} = \begin{pmatrix} \alpha_0 \begin{pmatrix} \beta_0 \\ \beta_1 \end{pmatrix} \\ \alpha_1 \begin{pmatrix} \beta_0 \\ \beta_1 \end{pmatrix} \end{pmatrix} = \begin{pmatrix} \alpha_0 \beta_0 \\ \alpha_0 \beta_1 \\ \alpha_1 \beta_0 \\ \alpha_1 \beta_1 \end{pmatrix}.$$

So we arrive at $|\psi\rangle = \alpha_0 \beta_0 |00\rangle + \alpha_0 \beta_1 |01\rangle + \alpha_1 \beta_0 |10\rangle + \alpha_1 \beta_1 |11\rangle$. One can quickly verify that $|\alpha_0 \beta_0|^2 + |\alpha_0 \beta_1|^2 + |\alpha_1 \beta_0|^2 + |\alpha_1 \beta_1|^2 = 1$ if and only if $|\alpha_0|^2 + |\alpha_1|^2 = 1$ and $|\beta_0|^2 + |\beta_1|^2 = 1$, that is if and only if $|\psi_0\rangle$ and $|\psi_1\rangle$ are both valid quantum states.

Example 2.6   It is important to note that *not* all multi-qubit states can be written in the tensor product form. The class of multi-qubit quantum states that cannot be expressed in terms of a tensor product of two quantum states is called the *entangled states*. One famous example is the Bell state:

$$|\psi\rangle = \frac{1}{\sqrt{2}}(|00\rangle + |11\rangle).$$

The key observation is that $|\psi\rangle \neq (\alpha_0 |0\rangle + \alpha_1 |1\rangle) \otimes (\beta_0 |0\rangle + \beta_1 |1\rangle)$ for any valid choices of $\alpha_0, \alpha_1, \beta_0$, and β_1. Quantum states that *can* be written in a tensor product of two states are called the *separable states* or *product states*.

Visualizing A Qubit: The Bloch Sphere

In order to visualize how operations on qubits affect the quantum state, we first need a geometric representation of the qubit. Recall we need two complex numbers α, β in order to represent the entire state of a qubit. Each complex number can be specified entirely by two real numbers, a, b. This means to represent the qubit we should need four dimensions. However, we are only able to visualize things in at most three dimensions. The Bloch Sphere coincides with the so-called *principle axes of spin measurement*, that is $\hat{x}, \hat{y}$, and $\hat{z}$. The Bloch sphere is to visualize a qubit in three dimensions. First, we rewrite the quantum state $|\psi\rangle$ into three real numbers (i.e., a, b, and φ), and then reduce to two (i.e., θ and φ) after applying the normalization condition:

$$|\psi\rangle = ae^{i\varphi}|0\rangle + be^{-i\varphi}|1\rangle = \cos\left(\frac{\theta}{2}\right)e^{i\varphi}|0\rangle + \sin\left(\frac{\theta}{2}\right)e^{-i\varphi}|1\rangle.$$

This equation now has only two unknowns φ and θ. This is enough to represent the qubit in three dimensions using spherical coordinates with a fixed radius $r = 1$. That is the quantum state $|\psi\rangle$ is a vector in $\mathbb{R}^3$ given by $(1, \theta, -\varphi)$. This can be visualized as a point on the surface of the Bloch sphere as follows:

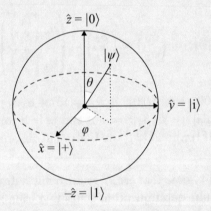

Equivalently, we can write a quantum state $|\psi\rangle$ in the Cartesian coordinate (x, y, z):

$$\rho = |\psi\rangle\langle\psi| = \frac{1}{2}(I + x\sigma_x + y\sigma_y + z\sigma_z),$$

where I is the identity matrix and σ_i are the Pauli matrices. The above formula is also known as the *Bloch sphere representation* of a quantum state.

We have found a one-to-one mapping from a (pure) single-qubit state $|\psi\rangle$ to a point on the surface of the Bloch sphere. Here "pure" is to distinguish from another class of quantum states called "mixed states." A mixed state is a probability distribution of several pure states, that is $\rho = \sum_i p_i |\psi_i\rangle\langle\psi_i|$. All mixed states live at the *inside* of the Bloch sphere. No quantum state lives outside of the Bloch sphere.

It is important to note that a single-qubit Hilbert space is a two-dimensional complex inner-product space, and the Bloch sphere visualizes a single-qubit state as a vector in $\mathbb{R}^3$. Two perpendicular vectors (i.e., having inner product equals zero) point in the opposite direction, such as the $\hat{z} = |0\rangle$ and $-\hat{z} = |1\rangle$ in the figure.

2.2.3   MEASUREMENTS

How much information can we store in or get out of a single qubit? The amplitudes of a qubit state $|\psi\rangle = \alpha |0\rangle + \beta |1\rangle$ take complex coefficients. So there are infinite different states for just a single qubit. Can we possibly encode/decode infinite amount of information in one qubit then? Not so fast. The *quantum measurement postulate* in quantum mechanics states that the only way to read out information from a quantum system is by interacting with the system via measurement, from which we obtain a probabilistic outcome. Formally, we can define the following process.

Definition 2.7   (Measurement).   When we measure a qubit $|\psi\rangle = \alpha |0\rangle + \beta |1\rangle$ we observe the basis state $|0\rangle$ with probability $|\alpha|^2$ and the basis state $|1\rangle$ with probability $|\beta|^2$.

The process of measurement is irreversible and probabilistic, meaning once measurement has occurred, the state $|\psi\rangle$ *collapses* into one of the two basis states ($|0\rangle$ or $|1\rangle$) and the original quantum superposition cannot be recovered.

Example 2.8   If *MeasZ* is the measurement operator (along the computational axis), the measurement outcome for each qubit *MeasZ* $|\psi\rangle$ will be either $|0\rangle$ or $|1\rangle$ depending on its state. In the Bloch sphere picture, the *MeasZ* outcome is related to the *latitude* of the quantum state—global phase (longitude) does not matter (see Table 2.2).

Remark 2.9   (Measurement along arbitrary axis)   So far, all the examples are measurements along the z-axis. However, it is possible to measure along a different axis. To see that, let's first rephrase Definition 2.7. Observe $\langle 0|\psi\rangle = \begin{pmatrix} 1 & 0 \end{pmatrix} \begin{pmatrix} \alpha \\ \beta \end{pmatrix} = \alpha$. Similarly, we have $\langle 1|\psi\rangle = \beta$. Therefore, *MeasZ* gives value $|0\rangle$ with probability $|\langle 0|\psi\rangle|^2$ and value $|1\rangle$ with probability $|\langle 1|\psi\rangle|^2$. In general, suppose we have a set of *orthonormal* basis $B = \{|b_i\rangle\}$. Here, orthonormal means that $\forall i, j$, we have $\langle b_i|b_j\rangle = 1$ if $i = j$, and 0 otherwise. There exists a measurement operator M along that basis, where we obtain measurement outcome $|b_i\rangle$ with probability $|\langle b_i|\psi\rangle|^2$ for each i (see Table 2.3).

In practice, we can accomplish measurement along a different axis than the computational axis by applying a change-of-basis transformation U and then measure in the computational axis just as before. For example, measuring along the x-axis (i.e., in the $\{|+\rangle, |-\rangle\}$ basis), denoted as *MeasX*, can be accomplished with Hadamard transformations H and z-axis measurements *MeasZ*.

Table 2.2: Example measurement outcomes by *MeasZ* on initial state $|\psi\rangle$.

Initial State	Readout	Final State	Probability
$|\psi\rangle = |0\rangle$	0	$|0\rangle$	100%
$|\psi\rangle = |1\rangle$	1	$|1\rangle$	100%
$|\psi\rangle = |+\rangle$	0	$|0\rangle$	50%
	1	$|1\rangle$	50%
$|\psi\rangle = |-\rangle$	0	$|0\rangle$	50%
	1	$|1\rangle$	50%
$|\psi\rangle = \frac{1}{\sqrt{2}}|00\rangle + \frac{1}{\sqrt{2}}|11\rangle$	00	$|00\rangle$	50%
	11	$|11\rangle$	50%
$|\psi\rangle = \alpha|00\rangle + \beta|01\rangle + \gamma|10\rangle + \delta|11\rangle$	00	$|00\rangle$	$|\alpha|^2$
	01	$|01\rangle$	$|\beta|^2$
	10	$|10\rangle$	$|\gamma|^2$
	11	$|11\rangle$	$|\delta|^2$

Table 2.3: Example measurement outcomes by a non-computational basis on initial state $|\psi\rangle$.

Measurement Basis	Initial State	Readout	Final State	Probability
$\{|+\rangle, |-\rangle\}$	$|\psi\rangle = |+\rangle$	+	$|+\rangle$	100%
$\{|+\rangle, |-\rangle\}$	$|\psi\rangle = |-\rangle$	−	$|-\rangle$	100%
$\{|+\rangle, |-\rangle\}$	$|\psi\rangle = |0\rangle$	+	$|+\rangle$	50%
		−	$|-\rangle$	50%
$\{|+\rangle, |-\rangle\}$	$|\psi\rangle = |1\rangle$	+	$|+\rangle$	50%
		−	$|-\rangle$	50%
$\{|b_i\rangle\}_{i=0}^{d-1*}$	$|\psi\rangle$	b_i	$|b_i\rangle$	$|\langle b_i|\psi\rangle|^2$

For completeness, we describe the *general measurement rules* for (pure) quantum states. To start with, we pick a measurement basis set, which can be written as a set of matrices $\{M_i\}_i$ satisfying the completeness condition

$$\sum_i M_i^\dagger M_i = I.$$

For instance, for the computational basis measurement, we take $M_1 = |0\rangle \langle 0|$ and $M_2 = |1\rangle \langle 1|$. Upon measurement, we obtain the outcome "i" with probability

$$\mathbf{Pr}[\text{observe } i] = |M_i \,|\psi\rangle \,|^2 = \langle \psi | M_i^\dagger M_i |\psi\rangle \,,$$

which results in a quantum state

$$|\psi'\rangle = \frac{M_i \,|\psi\rangle}{|M_i \,|\psi\rangle \,|} = \frac{M_i \,|\psi\rangle}{\sqrt{\langle \psi | M_i^\dagger M_i |\psi\rangle}}.$$

2.2.4   QUANTUM GATES

But what are these transformations after all? We are going to formally introduce the *quantum gate postulate*, that is, how do we manipulate quantum states for computation. What kind of quantum logic operations can we achieve? How do we transform from one quantum state to another? Mathematically, this process is defined as a "norm-preserving linear transformation," in other words, a unitary transformation. Transforming from $|\psi\rangle = \alpha \,|0\rangle + \beta \,|1\rangle$ to $|\varphi\rangle = \alpha' \,|0\rangle + \beta' \,|1\rangle$, we must have $|\alpha|^2 + |\beta|^2 = |\alpha'|^2 + |\beta'|^2 = 1$ to ensure that both $|\psi\rangle$ and $|\varphi\rangle$ are valid quantum states. If we represent a quantum state $|\psi\rangle$ as a column vector as in Definition 2.1, then we can represent the quantum logic gate on the state vector by a linear operator U given by a matrix.

Definition 2.10   (Transformation).   A valid logical transformation must map a quantum state to another quantum state. That is, for $U : |\psi\rangle \to U \,|\psi\rangle$, we require $|\langle \psi | \psi \rangle|^2 = 1 = |\langle \psi | U^\dagger U |\psi\rangle|^2$. Formally, this means that U is represented by a *unitary matrix* (i.e., $U^\dagger U = I$).

Unlike measurement operators which are irreversible and probabilistic, such logical transformation is *reversible* (since unitary matrix U is always invertible) and *deterministic* (since U maps any fixed initial state $|\psi\rangle$ to a unique final state $U \,|\psi\rangle$). Physically, it means that the transformation is energetically coherent and we can always undo this process by inverting the action. From an information theory perspective, it means that no information is destroyed (or leaked to the environment) under unitary transformations. In other words, knowing the output and what transformation it underwent, we can always recover the input. Notice that this is not always the case in classical boolean logic. Take a common logic gate, the AND gate, as an example— knowing that we obtained the bit 0 from an AND operation of two bits x and y, i.e., AND $(x, y) = 0$, we cannot tell if we started with $(x, y) = (0, 0)$ or $(0, 1)$ or $(1, 0)$. Hence, we call the AND gate an irreversible gate. An example of nontrivial classical reversible gate is the NOT gate, which negates the two states 0 and 1. Transformations via quantum logic gates, however, are all reversible. It is important to point out that the transformation principle does not account for the effect of noise. For instance, a qubit, when perturbed by the environment, can decohere

Table 2.4: Example quantum gates and a selection of their algebraic properties.

Quantum Gate	Circuit Form	Matrix Form	Truth Table	Algebraic Properties
Identity gate (I)	$-\boxed{I}-$	$I = \begin{pmatrix} 1 & 0 \\ 0 & 1 \end{pmatrix}$	$\|0\rangle \mapsto \|0\rangle$ $\|1\rangle \mapsto \|1\rangle$	
Not gate (X)	$-\boxed{X}-$	$X = \begin{pmatrix} 0 & 1 \\ 1 & 0 \end{pmatrix}$	$\|0\rangle \mapsto \|1\rangle$ $\|1\rangle \mapsto \|0\rangle$	$X^2 = Y^2 = Z^2$ $= -iXYZ = I,$
Y gate (Y)	$-\boxed{Y}-$	$Y = \begin{pmatrix} 0 & -i \\ i & 0 \end{pmatrix}$	$\|0\rangle \mapsto \|i\rangle$ $\|1\rangle \mapsto \|-i\rangle$	$XY = -YX = iZ,$ $YZ = -ZY = iX,$
Z gate (Z)	$-\boxed{Z}-$	$Z = \begin{pmatrix} 1 & 0 \\ 0 & -1 \end{pmatrix}$	$\|0\rangle \mapsto \|0\rangle$ $\|1\rangle \mapsto -\|1\rangle$	$ZX = -XZ = iY.$
Phase gate (S)	$-\boxed{S}-$	$S = \begin{pmatrix} 1 & 0 \\ 0 & i \end{pmatrix}$	$\|0\rangle \mapsto \|0\rangle$ $\|1\rangle \mapsto i\|1\rangle$	$S^2 = Z$
T gate (T)	$-\boxed{T}-$	$S = \begin{pmatrix} 1 & 0 \\ 0 & e^{i\frac{\pi}{4}} \end{pmatrix}$	$\|0\rangle \mapsto \|0\rangle$ $\|1\rangle \mapsto e^{i\frac{\pi}{4}}\|1\rangle$	$T^2 = S$ $TXT^\dagger = e^{-i\pi/4} SX,$
Hadamard gate (H)	$-\boxed{H}-$	$H = \frac{1}{\sqrt{2}} \begin{pmatrix} 1 & 1 \\ 1 & -1 \end{pmatrix}$	$\|0\rangle \mapsto \|+\rangle$ $\|1\rangle \mapsto \|-\rangle$	$H^2 = I,$ $X = HZH.$

to a classical state. Such a process is incoherent and not reversible. We will defer the discussion on the effect of noise to Chapter 8. For simplicity, this chapter will assume an ideal, noise-free situation.

Example 2.11   Quantum logic gates define the set of elementary operations that we can perform in a quantum computer. Let's start with the simplest example, namely quantum gates on a single qubit. A single-qubit gate can be viewed as a transformation that takes one point on the Bloch sphere to another by rotating by an arbitrary angle along a certain axis. Table 2.4 shows a few examples of single-qubit operations.

For example, when a qubit is in a superposition state $|\psi\rangle = \alpha |0\rangle + \beta |1\rangle$ then the operation applies to each of the basis states, e.g.,

$$H |\psi\rangle = \alpha(H |0\rangle) + \beta(H |1\rangle) = \alpha |+\rangle + \beta |-\rangle = \frac{\alpha + \beta}{\sqrt{2}} |0\rangle + \frac{\alpha - \beta}{\sqrt{2}} |1\rangle.$$

X gate, Y gate, and Z gate are π (or 180°) rotations about the x-axis, y-axis, and z-axis of the Bloch sphere, respectively. S gate performs a $\frac{\pi}{2}$ rotation about the z-axis, thus we have $S^2 = Z$. T gate performs a $\frac{\pi}{4}$ rotation about the z-axis, thus $T^2 = S$. H (Hadamard) gate is a π rotation about an axis diagonal in the x–z plane.

Example 2.12   It is usually convenient to include generic single-qubit rotation gates (e.g., R_x, R_y, R_z gates) along the Pauli axes in our gate set. We write $R_x(\theta)$ to indicate a rotation of θ angle about the x-axis. Several of the gates we've already discussed are just examples of the $R_z(\theta)$ gates, specifically the Z, S, and T gates which rotate by a π, $\frac{\pi}{2}$, and $\frac{\pi}{4}$ angle, respectively. Formally, the rotation gate can be written in their matrix forms as follows:

$$R_x(\theta) = \cos\frac{\theta}{2}I - i\sin\frac{\theta}{2}X = \begin{pmatrix} \cos\frac{\theta}{2} & -i\sin\frac{\theta}{2} \\ -i\sin\frac{\theta}{2} & \cos\frac{\theta}{2} \end{pmatrix}$$

$$R_y(\theta) = \cos\frac{\theta}{2}I - i\sin\frac{\theta}{2}Y = \begin{pmatrix} \cos\frac{\theta}{2} & -\sin\frac{\theta}{2} \\ \sin\frac{\theta}{2} & \cos\frac{\theta}{2} \end{pmatrix}$$

$$R_z(\theta) = \cos\frac{\theta}{2}I - i\sin\frac{\theta}{2}Z = \begin{pmatrix} e^{-i\frac{\theta}{2}} & 0 \\ 0 & e^{i\frac{\theta}{2}} \end{pmatrix}.$$

Example 2.13   **Two-qubit gates** take two qubits as inputs. They typically have an "entangling" effect—the operation applied to one qubit is dependent on the state of the other qubit, in other words, they are conditional gates. Among the most common two-qubit operations are the controlled-not gate (or CNOT gate), and the controlled-phase gate (or CZ gate), as shown in Table 2.5.

In the example, the CNOT gate is a two-input two-output gate which performs a NOT operation on the second (target) qubit only when the first (control) qubit is $|1\rangle$. Similarly for CZ gate, if the control qubit is $|1\rangle$, then we apply a Z gate to the target qubit. But looking at the truth table of the CZ gate, we notice that, in fact, it makes no distinction between the first and the second qubits—a phase is accumulated for the $|11\rangle$ basis. Hence, the CZ gate has a symmetric circuit symbol. One can in fact implement a CNOT gate with a CZ gate and vice versa. For example, CNOT is equivalent to a CZ gate with two Hadamard gates on both sides, since $HZH = X$:

The fact that these gates are conditional gates can also be observed from their matrix representations. In general, we may construct a controlled version of any gate U. Notice that

Table 2.5: Example measurement outcomes by $MeasZ$ on initial state $|\psi\rangle$.

Quantum Gate	Circuit Form	Matrix Form	Truth Table
CNOT gate		$CNOT = \begin{pmatrix} 1 & 0 & 0 & 0 \\ 0 & 1 & 0 & 0 \\ 0 & 0 & 0 & 1 \\ 0 & 0 & 1 & 0 \end{pmatrix}$	$\begin{aligned} &\|00\rangle \mapsto \|00\rangle \\ &\|01\rangle \mapsto \|01\rangle \\ &\|10\rangle \mapsto \|11\rangle \\ &\|11\rangle \mapsto \|10\rangle \end{aligned}$
CZ gate		$CZ = \begin{pmatrix} 1 & 0 & 0 & 0 \\ 0 & 1 & 0 & 0 \\ 0 & 0 & 1 & 0 \\ 0 & 0 & 0 & -1 \end{pmatrix}$	$\begin{aligned} &\|00\rangle \mapsto \|00\rangle \\ &\|01\rangle \mapsto \|01\rangle \\ &\|10\rangle \mapsto \|10\rangle \\ &\|11\rangle \mapsto -\|11\rangle \end{aligned}$

controlled-U gate can be written as the sum of two terms, namely, when the first qubit is $|0\rangle$, nothing happens to the second qubit, and when the first qubit is $|1\rangle$, then we apply U gate on the second qubit:

$$\text{controlled-}U = \Lambda(U) = |0\rangle\langle 0| \otimes I + |1\rangle\langle 1| \otimes U = \left(\begin{array}{c|c} I & 0 \\ \hline 0 & U \end{array}\right)$$

where the notation $\Lambda(\cdot)$ stands for a controlled version of a gate. One can quickly verify that these controlled gates usually have an *entangling* effect. In particular, they can transform a product state as input into an entangled state as output. For example, the following circuit produces the Bell state:

$$\begin{array}{c} |0\rangle \; \boxed{H} \bullet \\ |0\rangle \quad\;\; \oplus \end{array} \Bigg\} \frac{1}{\sqrt{2}}(|00\rangle + |11\rangle)$$

Another gate, important in architectures which require qubits to be adjacent in order to perform multi-qubit operations, is the SWAP gate, which switches the states of two qubits, which is equivalent to interleaving three CNOT gates:

$$\times \quad = \quad \oplus \bullet \oplus$$

It can be shown that single qubit gates and two qubit gates are *universal* for arbitrary quantum logic. In other words, any unitary gate on multiple qubits can be decomposed into a sequence of one and two-qubit gates. One example of a universal gate set found commonly in the literature is

$$\mathcal{G} = \{H, T, CNOT\}.$$

Physically, realizing a multi-qubit gate is extremely challenging. So finding an efficient decomposition of a unitary gate into a sequence of smaller unitary gates from a chosen gate set is critical to the success of executing a quantum circuit. This problem is often referred to as *quantum compilation*. We will revisit exactly this problem but in much greater detail in Chapter 6.

Example 2.14   Three-qubit gates. These gates may be controlled on more than one qubit. One of the most famous examples is the Toffoli Gate (CCNOT or Controlled-Controlled-Not). It has the following circuit:

The Toffoli gate can be used to achieve irreversible classical operations like AND and OR in quantum computing in a reversible manner.

2.3   NOISY QUANTUM SYSTEMS

How do quantum systems interact with the environment? How do we characterize a quantum process in the presence of noise? In this section, we extend our discussion to include noisy quantum systems.

To begin with, we emphasize that a quantum system is inherently probabilistic—when implemented in practice, quantum systems have to involve some incoherence processes, whether they are intentional (e.g., by measurements) or unintentional (e.g., by random perturbations), and they ultimately lead to probabilistic outcomes. For example, an ideal system can prepare a quantum state $|\psi\rangle$ with certainty using a perfect quantum circuit, while a noisy quantum system likely produces a random distribution of quantum states, i.e., $|\psi_i\rangle$ with probability p_i, due to imprecise controls or environmental noise. To model these effects, we need a more general definition of a probabilistic quantum state.

2.3.1   QUANTUM PROBABILITY

Recall that upon measurement of a quantum state $|\psi\rangle = \sum_i \alpha_i |x_i\rangle$, we obtain a classical probability distribution over the measurement outcomes $\{|x_i\rangle\}_i$, according to the probability $\{|\alpha_i|^2\}_i$. Indeed, the sum of probability equals 1, i.e., $\langle \psi | \psi \rangle = \sum_i |x_i|^2 = 1$. We can rewrite the probability expression for the outcome x_i as

$$\mathbf{Pr}[Meas(|\psi\rangle) = |x_i\rangle] = |\alpha_i|^2 = |\langle \psi | x_i \rangle|^2 = \langle \psi | x_i \rangle \langle x_i | \psi \rangle.$$

Let us denote $\Pi_{x_i} = |x_i\rangle \langle x_i|$ as a *projector*, as the effect of Π_{x_i} is a projection onto the subspace spanned by $|x_i\rangle$. Note that Π is a projection if it can be written as $\Pi = \sum_k |v_k\rangle \langle v_k|$ where v_k's are orthonormal (i.e., $\langle v_i|v_j\rangle = \delta_{ij}$, where δ_{ij} is the Kronecker delta). Equivalently, Π is a positive semidefinite (PSD) matrix such that $\Pi^2 = \Pi$, i.e., projecting twice is identical to projecting once.

We want to find the right notions of a noisy quantum state. We start by modeling it as a probability distribution over pure quantum states, $\{p_i, |\psi_i\rangle\}_i$.

Definition 2.15   The *mixed state* $\{p_i, |\psi_i\rangle\}$ of a quantum system is represented by the matrix

$$\rho = \sum_i p_i |\psi_i\rangle \langle \psi_i|.$$

This is called the *density matrix representation* of a quantum state. Furthermore, if ρ represent a mixed quantum state, it must satisfy $\mathrm{tr}(\rho) = 1$, and ρ is positive semidefinite (PSD), where $\mathrm{tr}(\cdot)$ is the trace of a matrix. Upon measuring this mixed state ρ, we obtain

$$\mathbf{Pr}[\mathrm{Meas}(\rho) = |x_i\rangle] = \sum_j p_j \, \mathbf{Pr}[\mathrm{Meas}(|\psi_j\rangle = |x_i\rangle)]$$

$$= \sum_j p_j \langle \psi_j|\Pi_{x_i}|\psi_j\rangle$$

$$= \sum_j p_j \mathrm{tr}(|\psi_j\rangle \langle \psi_j| \, \Pi_{x_i})$$

$$= \mathrm{tr}(\rho \Pi_{x_i}).$$

Let us now extend the measurement rules in Section 2.2.3 to mixed states. Again, we start with a set of measurement operators $\{M_i\}_i$ satisfying the completeness condition $\sum_i M_i^\dagger M_i = I$. We obtain the measurement outcome "i" with probability

$$\mathbf{Pr}[\text{observe } i] = \mathrm{tr}(\rho M_i^\dagger M_i) = \mathrm{tr}(M_i \rho M_i^\dagger),$$

where we used the property $\mathrm{tr}(AB) = \mathrm{tr}(BA)$. The resulting mixed quantum state is

$$\rho' = \frac{M_i \rho M_i^\dagger}{\mathrm{tr}(M_i \rho M_i^\dagger)}.$$

To tie closely with the notions of classical probability, we can define the measurement process as *observables*, and model the outcomes using the expectations of the observables. In particular, suppose we perform a measurement $\{M_1, \ldots, M_k\}$ on a quantum state ρ. We report a value λ_i if measurement outcome i is received. This is denoted as an observable

$\mathcal{O} = \lambda_1 M_1 + \lambda_2 M_2 + \cdots + \lambda_k M_k$. As a result, we obtain a *random variable* $\mathbf{x}$ that takes value λ_i with probability $\text{tr}(\rho M_i^\dagger M_i)$. We call the *expectation* of the observable $\mathcal{O}$ with respect to state ρ as

$$\mathbf{E}[\mathcal{O}] = \mathbf{E}[\mathbf{x}] = \text{tr}(\rho \mathcal{O}).$$

Now we continue to discuss the resulting mixed state when a pure quantum state is measured *partially*. Suppose an n-qubit quantum state is shared among two parties, e.g., Alice holds on to the first $n/2$ qubits and Bob holds on to the rest. It is then natural to ask: what is the state of Alice's qubits, if Bob measured his qubits and obtained a probabilistic outcome?

Definition 2.16 Given a bipartite quantum system in the form of a $d^2 \times d^2$ matrix $A \otimes B$ (where $d = 2^{n/2}$, A, and B are $d \times d$ matrices), the *partial trace* (over B) of the system is defined as

$$\text{tr}_B(A \otimes B) = A \cdot \text{tr}(B).$$

For a generic pure state (possibly entangled between A and B) in the density matrix form

$$\rho = |\psi\rangle \langle\psi| = \left(\sum_{i_A, i_B} \alpha_{i_A, i_B} |i_A\rangle \otimes |i_B\rangle \right) \left(\sum_{j_A, j_B} \alpha_{j_A, j_B} |j_A\rangle \otimes |j_B\rangle \right)^\dagger$$

$$= \sum_{i_A, i_B, j_A, j_B} \alpha_{i_A, i_B} \alpha_{j_A, j_B}^* |i_A\rangle \langle j_A| \otimes |i_B\rangle \langle j_B|.$$

After Bob measures his qubits, Alice's state becomes the partial trace of the quantum state over Bob's subsystem:

$$\rho_A = \text{tr}_B(|\psi\rangle \langle\psi|)$$

$$= \sum_{i_A, i_B, j_A, j_B} \alpha_{i_A, i_B} \alpha_{j_A, j_B}^* |i_A\rangle \langle j_A| (\text{tr}(|i_B\rangle \langle j_B|))$$

$$= \sum_m \sum_{i_A, j_A} \alpha_{i_A, m} \alpha_{j_A, m}^* |i_A\rangle \langle j_A|.$$

To quantitatively study the impact of noise, we need distance measures between quantum states or quantum processes. We postpone the technical details to Section 8.1 where we discuss the noise mitigation strategies.

2.3.2 OPERATOR SUM REPRESENTATION

Our goal in this section is to model the interaction between a quantum system with the environment, and thus describe the impact of noise on a quantum state. Fortunately, we already have all the tools we need: namely unitary transformation and measurement. In particular, we can

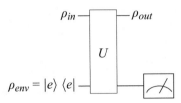

Figure 2.5: The unitary coupling picture between the systems and the environment.

view the impact of environmental noise in this following "unitary coupling evolution" picture, as shown in Figure 2.5.

A unitary transformation U is applied to both the environment and the system ($\rho_{env} \otimes \rho_{in}$) (Figure 2.5), followed by an implicit partial measurement over the environment. Notice that we write environment first in the tensor product, for the sake of convenience in later discussions.

More generally, any physical processes that can happen to a mixed quantum state can be written as a *linear map*:

$$\rho \to \mathcal{E}(\rho).$$

Such a linear map is sometimes referred to as a *superoperator*. For example, a unitary transformation (without interaction with the environment) can be written as an unitary operator: $\mathcal{E}(\rho) = U\rho U^{\dagger}$.

The goal is to write down an operator form for the entire unitary coupling evolution (which involves the ρ_{env}, a unitary transformation U, and a measurement):

$$\rho_{in} \to \rho_{out} = \text{tr}_{env}(U(\rho_{env} \otimes \rho_{in})U^{\dagger}).$$

Here U acts on both the system and the environment. Suppose we start with $\rho_{env} = |e\rangle \langle e|$, and arbitrary measurement operators $M_k = |e_k\rangle \langle e_k|$, such that $|e_k\rangle$'s form an orthonormal basis for the space of the environment. Then we have

$$\mathcal{E}(\rho) = \text{tr}_{env}\left(U(\rho_{env} \otimes \rho)U^{\dagger}\right)$$
$$= \sum_{k} \langle e_k| U (|e\rangle \langle e| \otimes \rho) U^{\dagger} |e_k\rangle.$$

The key step is to define an operator $E_k = \langle e_k|U|e\rangle$. Intuitively, we take the unitary U (acting on both the system and the environment) and cut it into separate operators E_k, each acting on just the system:

$$U(|e\rangle \otimes |\psi\rangle) = \sum_{k} |e_k\rangle \otimes E_k |\psi\rangle.$$

Therefore, the overall linear map can be rewritten in terms of the operators:

$$\mathcal{E}(\rho) = \sum_{k} E_k \rho E_k^{\dagger}.$$

This is called the *operator sum representation* (OSR) of a quantum process [85], and the E_k's are commonly named the *Kraus operators*. The condition for a set of valid Kraus operators is $\sum_k E_k^\dagger E_k = I$. Notice that the operators often are non-unitary matrices. It is also important to note that the OSR is non-unique for a given linear map $\epsilon(\rho)$, because one can convert to another set of operators, F_k, by changing basis in the measurements, M_k, for the environment.

2.3.3   QUBIT DECOHERENCE AND GATE NOISE

In this section, we use the operator sum representation to model some examples of quantum noise, namely the amplitude damping noise, the phase damping noise, and the depolarizing noise.

Definition 2.17   An *amplitude damping noise* can be represented as

$$\mathcal{E}(\rho) = E_0 \rho E_0^\dagger + E_1 \rho E_1^\dagger,$$

where $E_0 = \begin{pmatrix} 1 & 0 \\ 0 & \sqrt{1-\gamma} \end{pmatrix}$ and $E_1 = \begin{pmatrix} 0 & \sqrt{\gamma} \\ 0 & 0 \end{pmatrix}$, for some γ parameter between 0 and 1.

This model captures the general behavior of a quantum system losing energy. For instance, the spontaneous emission of electromagnetic radiation for an atom can be modeled as amplitude damping, where γ is the probability of emission. The effect of energy loss can be seen from the fact that E_1 brings the amplitude on $|1\rangle$ (excited state) to $|0\rangle$ (ground state).

In a more realistic setting, the parameter γ is a time-dependent function, which is often characterized by $1 - e^{t/T_1}$, where t is time and T_1 is called the "spin-lattice relaxation time" or the "T_1 coherence time." As time goes by, a quantum state is, therefore, exponentially more likely to undergo energy loss, and T_1 is a parameter characterizing the speed of such process. More specifically, the generalized version of amplitude damping describes the T_1 relaxation process, where the Kraus operators are:

$$E_0 = \sqrt{p} \begin{pmatrix} 1 & 0 \\ 0 & \sqrt{1-\gamma} \end{pmatrix}, E_1 = \sqrt{p} \begin{pmatrix} 0 & \sqrt{\gamma} \\ 0 & 0 \end{pmatrix},$$

$$E_2 = \sqrt{1-p} \begin{pmatrix} \sqrt{1-\gamma} & 0 \\ 0 & 1 \end{pmatrix}, E_3 = \sqrt{1-p} \begin{pmatrix} 0 & 0 \\ \sqrt{\gamma} & 0 \end{pmatrix},$$

where the state converges to the mixed state $\rho_\infty = \begin{pmatrix} p & 0 \\ 0 & 1-p \end{pmatrix}$.

Definition 2.18   A *phase damping noise* can be represented as

$$\mathcal{E}(\rho) = E_0 \rho E_0^\dagger + E_1 \rho E_1^\dagger,$$

where $E_0 = \begin{pmatrix} 1 & 0 \\ 0 & \sqrt{1-\lambda} \end{pmatrix}$ and $E_1 = \begin{pmatrix} 0 & 0 \\ 0 & \sqrt{\lambda} \end{pmatrix}$, for some λ parameter between 0 and 1.

In a realistic setting, phase damping is thought to be related to the loss of quantum information without any loss of energy. Equivalently, it is a process where the qubit undergoes a phase flip (i.e., Z gate) with probability $(1 - \sqrt{\lambda})/2$. Similarly, the parameter λ in phase damping is often characterized by a time-dependent function $1 - e^{t/T_2}$, where T_2 is called the "spin-spin relaxation time" or the "T_2 coherence time." In general, since amplitude damping contributes to both T_1 and T_2 rates, for a system with both amplitude and phase damping, we have $T_1 \geq 2T_2$. It is also worth noting that amplitude damping contributes to both T_1 and T_2 relaxation [86]. To accurately capture the behavior of qubit decoherence, T_1 and T_2 are typically separately during idle or gate time, as qubits usually decohere faster during gate time.

Another commonly studied model for capturing gate noise is the depolarizing noise (or sometimes referred to as a special case of stochastic Pauli noise).

Definition 2.19   A *depolarizing noise* can be represented as

$$\mathcal{E}(\rho) = (1 - p)I\rho I + \frac{p}{3}X\rho X + \frac{p}{3}Y\rho Y + \frac{p}{3}Z\rho Z,$$

so the corresponding Kraus operators are $\{\sqrt{1-p}I, \sqrt{p/3}X, \sqrt{p/3}Y, \sqrt{p/3}Z\}$.

This can be interpreted as a process where the state ρ is unchanged with probability $1 - p$ and applied with X, Y and Z with equal probability $p/3$. Due to the observation that $I/2 = (\rho + X\rho X + Y\rho Y + Z\rho Z)/4$ for arbitrary ρ, we can rewrite

$$\mathcal{E}(\rho) = \left(1 - \frac{4p}{3}\right)\rho + \frac{4p}{3}\frac{I}{2},$$

which can be equivalently interpreted as the quantum state is unchanged with probability $1 - \frac{4p}{3}$, and replaced by $\frac{I}{2}$ with probability $\frac{4p}{3}$.

So far we have seen several simple noise models; to realistically characterize a noisy quantum system, we need more sophisticated models than the ones introduced here. Please see Chapter 8 for more details.

2.4   QUBIT TECHNOLOGIES

This section is a computer scientist's guide to the basics of qubit technologies. But why does a computer scientist need the technical know-how in the first place? After all, in today's classical computing community, few programmers need any knowledge of how transistors work. Quantum computers require strict isolation and coherent manipulation of complex, physical systems to a level of precision never before attempted.

Although there already exists an ecosystem of layered quantum software tools and abstractions that serve as an interface between those layers, it is perhaps premature and fallacious to follow a model too similar to classical software. Some existing algorithms and systems tools are unrealistic in the short term because they were developed with idealistic assumptions about the underlying hardware. A decent appreciation of how qubits behave will come in handy when designing more efficient algorithms and software systems. For this reason, we dedicate the rest of the chapter to a gentle introduction to the leading technologies for realizing quantum computing devices.

Today, experimentalists are building qubit systems in carefully controlled laboratory environments. The leading technologies that may have the potential for realizing scalable quantum computing include trapped ion qubits, superconducting qubits, semiconductor spin qubits, linear optics, and Marjorana qubits, etc. The general philosophy of qubit design can be summarized in the *DiVincenzo Criteria* [87]:

1. scalable system with well-characterized qubits;

2. ability to initialize qubits (e.g., prepare in computational basis);

3. stability of qubits (i.e., long decoherence times);

4. support for a universal instruction set (e.g., single qubit gates and CNOT gate) for arbitrary computation; and

5. ability to measure qubits (e.g., readout in computational basis).

Note that these goals are in tension with each other. In particular, being able to initialize, perform gates, and measure requires interactions between the system and environment, but long decoherence times require isolating the system from the environment. This is the fundamentally difficult part about building a quantum computer.

Tremendous progress has been made over the past few decades. A wide range of physical systems have shown to have the potential to implement qubits, and some have been demonstrated with proposals for scalable architectures. In the following, we describe two important technologies that have attracted the most interests in research labs, large companies, and startups.

2.4.1   TRAPPED ION QUBITS

One of the most natural ways of making a qubit is to use an atomic ion. An atomic ion makes a great qubit because its internal energy levels exhibit quantum mechanical properties. In the following, we introduce the basics of making trapped ion qubits, and how they can be integrated into a quantum computing system.

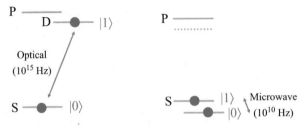

Figure 2.6: State transitions for two common types of trapped ion qubits: the optical qubit and the hyperfine qubit.

Types of Atomic Ion Qubits

In an atomic ion qubit, the quantum states are represented by two internal energy levels of the ion (that is, one for $|0\rangle$ and the other for $|1\rangle$). An atomic ion (such as Ca^+, Sr^+, Ba^+, and Yb^+) typically has more than two internal energy levels. So the choice of two of the levels determines how the qubit is controlled. The following two leading designs pick the energy levels differently, as shown in Figure 2.6.

- *Optical Qubits.* On the left, we have an optical qubit because the two chosen energy levels have a separation of about 10^{15} Hz, which is around the frequency of visible light. The two energy levels are from different orbitals of the ion, $|0\rangle$ from s orbital and $|1\rangle$ from d orbital. If the frequency of the laser beam matches the transition frequency from $|0\rangle$ to $|1\rangle$, the ion is excited after absorbing energy from the laser. The ion also makes spontaneous decay from the excited energy state to the lower energy state in around 1 second. Common ions that can be made into optical qubits include Ca^+, Sr^+, Ba^+, and Yb^+.

- *Hyperfine Qubits.* In contrast, a hyperfine qubit on the right chooses both energy levels from the s orbital, and thus has an energy separation of about 10^{10} Hz, which falls in the microwave spectrum [88]. A hyperfine qubit can be directly driven via microwave control or via Raman transitions which we will show in greater detail below. Ions commonly made into hyperfine qubits include Ca^+, Sr^+, Ba^+, and Yb^+, Be^+, Mg^+, Hg^+, Cd^+, and Zn^+. Throughout the rest of the section, we use $^{171}Yb^+$ hyperfine qubit as example [89].

Once the qubit states are defined, we need to know how to perform measurement and apply high-fidelity single- and two-qubit gates on them.

Measuring a Qubit

The measurement of a trapped ion hyperfine qubit is achieved via *state-dependent fluorescence* [88, 90, 91]. As Figure 2.7 shows, an optical drive is carefully tuned to match a transition from the

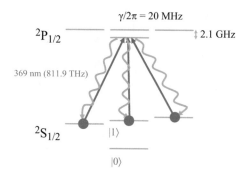
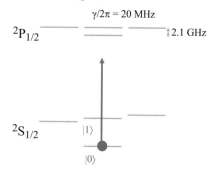

Figure 2.7: Measurement outcome is observed by state-dependent flourescence.

$|1\rangle$ state to an energy level in the p orbital, so that a spontaneous decay from the p orbital back to s orbital will emit photons, lighting up the ion. In fact, it is slightly detuned from the p level such that the decay happens instantaneously. However, if the qubit is originally in the $|0\rangle$ state, such transition will not happen. The $^2s_{1/2} \rightarrow^2 p_{1/2}$ is a cycling transition, which means that if we continue applying the Raman beams, the ions will remain fluorescent.

Single-Qubit Gate: Raman or Microwave Transition

In a trapped ion quantum computer, qubits are controlled by carefully tuned laser beams. When the frequency of the laser matches the separation between two states, population in the lower state will be excited to the higher state. For a hyperfine qubit, qubit states have separation for around 10^{10} Hz, so state transitions can be controlled directly via microwave pulses, as shown on the left of Figure 2.8. The advantage of microwave-controlled single-qubit gate is its low error rates (10^{-6}), while the disadvantage lies at its difficulty in focusing on individual ions due to its large wavelength. Alternatively, state transitions can happen via Raman transitions, that is, first exciting to a p state then decaying back to s states, as shown on the right of Figure 2.8. Again, the excitation is slightly detuned so that the spontaneous decay happens instantaneously. The Raman transition approach has slightly higher error rates (10^{-4}), but targets individual ions better. Arbitrary-angle single-qubit rotations $R_\phi(\theta)$ can be implemented by tuning the Raman beat-note. The angle θ and axis ϕ are determined by the duration and phase off-set of the Raman pulses.

Two-Qubit Gate: Ising (XX) Gate

The native two-qubit gate in a trapped ion system is called the XX gate or Ising gate [92–95]. Its entangling interaction is achieved via dipole-dipole coupling between two ions. A detuned Raman transition can apply spin-dependent forces on the ions, which triggers their motional excitations. The number of different modes multiple ions can move to is huge. Take an example where the ions physically oscillate in the direction perpendicular to the ion chain. On one hand,

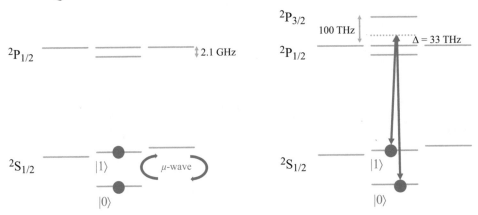

Figure 2.8: Single qubit gates via Raman transition or microwave transition.

if the two ions oscillate in phase (that is both moving synchronously up and down) then the distance between them is unchanged. On the other hand, the two ions oscillate out of phase (that is one going up and the other going down and vice versa) then the Coulomb force between them changes because their distance changes. We thus have a force that is dependent on the state of the ions. The motional excitations have an entangling effect because they lead to conditional phase shifts of the ions. Pulse shaping techniques are applied to disentangle the motions at the end of the gate. In particular, if the two ions are distance r apart and the oscillation is about δ, then the dipole-dipole coupling leads to a conditional phase shift of $\varphi = \frac{\Delta E t}{\hbar}$, where $\Delta E \approx -\frac{(e\delta)^2}{2r^3}$. This effective Ising interaction between the ions adds phase shifts depending on the spin of the ion:

$$
\begin{aligned}
|00\rangle &\mapsto |00\rangle \\
|01\rangle &\mapsto e^{-i\varphi} |01\rangle \\
|10\rangle &\mapsto e^{-i\varphi} |10\rangle \\
|11\rangle &\mapsto |11\rangle
\end{aligned}
$$

This is called an *XX* interaction, because its operator has the form of a $\sigma_x \otimes \sigma_x$ in the exponent:

$$
XX[\varphi] = e^{-i\sigma_x^{(1)} \otimes \sigma_x^{(2)} \varphi} = \begin{pmatrix} \cos(\varphi) & 0 & 0 & -i\sin(\varphi) \\ 0 & \cos(\varphi) & -i\sin(\varphi) & 0 \\ 0 & -i\sin(\varphi) & \cos(\varphi) & 0 \\ -i\sin(\varphi) & 0 & 0 & \cos(\varphi) \end{pmatrix}.
$$

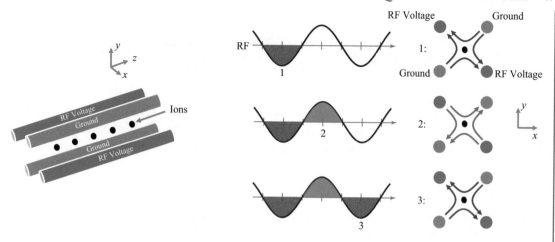

Figure 2.9: Schematics for a RF Paul trap.

This *XX* interaction can be used to implement the well-known *Mølmer-Sørensen gate*. For completeness, we show that the CNOT gate can be implemented using the following circuit:

$$
\begin{array}{c}
q_0 \\
q_1
\end{array}
\;=\;
R_y\!\left(\alpha\tfrac{\pi}{2}\right)\;XX\!\left(\alpha\tfrac{\pi}{4}\right)\;R_y\!\left(-\alpha\tfrac{\pi}{2}\right)\;R_z\!\left(-\tfrac{\pi}{2}\right)\;R_x\!\left(-\tfrac{\pi}{2}\right)
$$

Here the geometric phase of the XX gate is $\chi = \pm\tfrac{\pi}{4}$, and $\alpha = \mathrm{sgn}(\chi)$.

Loading (Trapping) Qubits

In this section, we describe how ions are being trapped in place and prepared to their initial states. In particular, one of the reasons that ions are chosen as qubits is because they are charged particles, which can feel the forces exerted on them by electromagnetic fields. However, it is not possible to create a field that exerts inward forces in all directions, as the number of electromagnetic field lines going into an enclosed system must equal to the number of lines going out. The best we can do is to create a stable equilibrium in one direction:

The *RF (radio-frequency) Paul trap* [96, 97] cleverly gets around this issue by applying a sinusoidal electric field quadruple around the ions to keep them stationary, as shown in Figure 2.9.

On the left, four rods of electrodes are shown. Ions are trapped in between the four rods, lining up in a chain. The voltages on the rods are applied (as shown on the right) so that the chain of ions at the center can be held (on average) stationary.

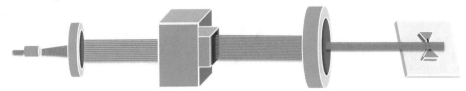

Figure 2.10: Schematics for a trapped ion QPU. After initialized from the optical source on the left, the laser are split into independently modulated beams, and then focused on the HOA trap on the right, providing individual controls over the array of ions in the trap.

Trapped Ion Quantum Processing Unit

Integrating the aforementioned key components into a system [98, 99], a trapped-ion QPU (quantum processing unit), such as the HOA (High Optical Access) trap [100], shown in Figure 2.10 is designed.

2.4.2   SUPERCONDUCTING QUBITS

In contrast to a trapped ion qubit, a superconducting qubit is implemented with macroscopic, lithographically printed circuit elements. The circuit elements are parameterized and configured such that they exhibit atom-like energy spectra, hence making an "artificial atom" with desired quantum mechanical properties. This technology has attracted significant industrial attention because it allows convenient design of qubits using existing integrated circuit (IC) technology. In the following, we introduce different designs of superconducting qubits, configured to operate in various regimes.

Superconducting Quantum Circuits

The most distinguishable element in a superconducting qubit is an electric circuit element called the *Josephson junction* [101, 102]. It is an insulator sandwiched between two superconductors. Below its critical temperature, a superconductor appears to have resistance dropped to zero, and pairs of electrons in the superconducting material form bonds, thus making them *Cooper pairs*. Ordinarily, single electrons have $\pm\frac{1}{2}$ spin; they are particles commonly referred to as fermions. But after forming as Cooper pairs, they have a total spin of 0 (making them particles with integer spins which are commonly referred to as bosons). With the superconductors in a Josephson junction, Cooper pairs can tunnel through the insulator in a quantized fashion (that is one pair at a time), thus giving rise to discrete energy levels needed for making a qubit. A superconducting qubit state is thus related to the number of Cooper pairs tunneled across the junction.

The Josephson junction element is used to implement what is known as an *anharmonic oscillator*. In contrast with a harmonic oscillator, where the energy levels are equally spaced by $\hbar\omega$, an anharmonic oscillator has unequal energy spacing. This "nonlinearity" is convenient because now we can drive the transitions between only two of the energy levels (usually the two lowest

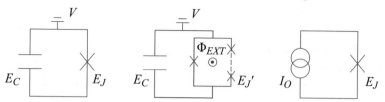

Figure 2.11: Types of superconducting qubits. **Left:** Circuit diagram for charge qubits (when $E_J \leq E_C$) and transmon qubit (when $E_J \gg E_C$), consisting of capacitor C and Josephson junction J. **Center:** Circuit diagram for a c-shunted flux qubit, where a junction is shunted with a number of junctions. **Right:** Circuit diagram for a phase qubit with current bias I_0.

levels) represented as qubit states for computation without exciting the other levels of the system. Normally, the higher the anharmonicity (i.e., difference between $\hbar\omega_{01}$ and $\hbar\omega_{12}$) the better we can individually address the computational states. However, in practice, anharmonicity also sets a limit on the speed of gate pulses we can apply to the qubit.

Types of Superconducting Qubits
Figure 2.11 are the common types of superconducting qubits.

- *Charge qubits.* A charge qubit defines its computational qubit state as the number of Cooper pairs on a superconducting island; this class includes the *Cooper-pair box* and the *transmon* qubit. Using the circuit shown in Figure 2.11, the superconducting island is located between one plate of the capacitor and the insulator of the Josephson junction. Operating in the "charge regime" (that is $E_J \leq E_C$), the qubit is controlled by a voltage source, which induces charge differences between the two sides of the Josephson junction. The qubit state $|0\rangle$ is given by the lack of Cooper pairs in the island, while $|1\rangle$ is given by the presence of a single Cooper pair. The charge qubit is also known as Cooper pair box [103–105]. The community has found that, in the charge regime, a qubit becomes highly sensitive to charge noise, making it hard to be kept coherent. Over time, more and more attention has been put on the flux regime (that is $E_J \gg E_C$) that trade charge noise for flux noise which appears to be more manageable. One can operate in the flux regime with $E_J \gg E_C$ (typically $E_J/E_C \geq 50$) by shunting the junction with a large capacitor, thus making $C_s \gg C_J$ and E_C small. This is commonly known as the transmon qubit [106].

- *Flux qubits.* In a flux qubit (as well as a fluxonium qubit [61, 107, 108]), the single Josephson junction is replaced by a SQUID (superconducting quantum interference device) [109]. A SQUID consists of a loop interupted by a number of Josephson junctions, where the effective critical current can be decreased by applying external magnetic flux Φ_{ext} through the loop. Thus, the effective E_J is tunable via changing the

SQUID's critical current using external flux. For a flux qubit, the energy levels correspond to the integer number of superconducting flux quanta induced in the loop: $\varphi_1 - \varphi_2 + \varphi_{ext} = 2\pi k$, where $\varphi_{ext} = \pi \Phi_{ext}/\Phi_0$ and $\Phi_0 = h/(2e)$ is a magnetic flux quantum. The discussions of superconducting qubits in the rest of the section will be centered around flux qubits as well as some of its variants, such as fluxonium [110].

- *Phase qubits.* A phase qubit [111, 112] defines its computational states using the quantum charge oscillation amplitudes across the Josephson junction, controlled with current biases. In contrast to a flux qubit, a phase qubit operates in a regime where $E_J/E_C \approx 10^6$.

Here we briefly describe how to perform measurement and elementary gate operations on superconducting qubits, following [113–115].

Measuring a Qubit

Qubit readout is typically performed via a technique called *dispersive readout* [116–119], which determines the qubit state via *state-dependent frequency shift* of a resonator coupled to each qubit. In the dispersive regime, where the detuning between the qubit and the resonator is large compared to their coupling rate, that is $|\omega_r - \omega_q| \ll g$, the qubit and the resonator push each other's frequencies with dispersive shifts. Since the shift on the resonator is state-dependent, we can use the changes in the frequency of the resonator to probe the state of the qubit without directly interacting with the qubit itself.

Single-Qubit Gate: Charge and Flux Drive

For transmon-like superconducting qubits, there are generally two classes of controls that drive individual qubits: (i) microwave control via a capacitively coupled resonator or feedline; it implements single-qubit rotations (along x axis and y axis); and (ii) flux control via external magnetic field; it implements z-axis single-qubit rotations or can be used to tune the frequency of qubits, as shown in Figure 2.12.

To enable microwave control, we couple the superconducting qubit to a microwave source (commonly referred to as charge drive or qubit drive) via a capacitor. The time-dependent voltage applied to the qubit can be written in a generic form

$$V_d(t) = V_0 v(t) = V_0(s(t)(\sin(\omega_d t)\cos(\phi) + \cos(\omega_d t)\sin(\phi))),$$

where V_0 is the pulse amplitude, $s(t)$ is a dimensionless (baseband) envelope function, ω_d is the driving frequency, ϕ is the phase offset determined arbitrarily, $\sin(\omega_d t)\cos(\phi) \equiv \sin(\omega_d t)I$ is the in-phase component of the pulse, and $\cos(\omega_d t)\sin(\phi)) \equiv \cos(\omega_d t)Q$ is the out-of-phase component of the pulse. Techniques like rotation wave approximation (RWA) can be used to show that, if the driving frequency equals the qubit frequency, the in-phase pulse corresponds to x-axis single-qubit rotations and the out-of-phase pulse corresponds to y-axis single-qubit

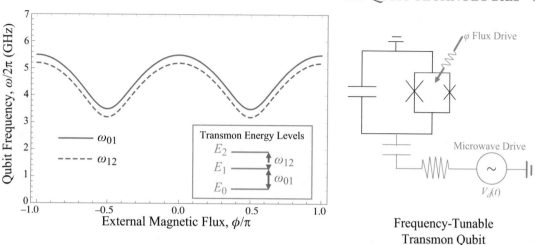

Figure 2.12: **Left:** Qubit frequencies as a function of external magnetic flux. The first three levels of the transmon, ω_{01} and ω_{12}, are plotted. **Right:** Circuit diagram for a frequency-tunable (asymmetric) transmon qubit (highlighted in black), consisting of a capacitor and two asymmetric Josephson junctions. Highlighted in gray are two control lines: the external magnetic flux control φ and microwave voltage drive line $V_d(t)$ for each transmon qubit.

rotations. More concretely, take an X gate as the example, we use a AWG (arbitrary waveform generator) to produce the pulse shape and send the signal in-phase through the qubit drive. The shape of the pulse is determined by the baseband $s(t)$ and the amplitude V_0, which can be derived by solving for the total phase gained during time t. Similarly for Y control, we solve for the pulse shape and send the signal out-of-phase through the qubit drive. Details are omitted here; we refer the interested reader to the tutorial in [113].

The choice of phase offset ϕ is arbitrary. Suppose we set $\phi \leftarrow \phi + \pi/2$, then the in-phase and out-of-phase components are swapped (up to change of sign)—I becomes Q and Q becomes $-I$. Recall from Section 2.2 that $ZX = iY$ and $ZY = -iX$. So shifting a phase in the AWG is equivalent to applying a Z gate. This way of implementing a Z gate by shifting the phase of subsequent pulses is called the *virtual Z* strategy [120].

Two-Qubit Gate: Flux or Microwave Control

Typical native two-qubit gates in a superconducting architecture include the iSWAP gate [60, 121–124] (for flux-tunable transmons), CZ (controlled-phase) gate [61, 125] (also for flux-tunable transmons), and CR (cross-resonance) gate [126–129].

For two flux-tunable transmons (coupled via a capacitor), the strategy for interacting two qubits is by tuning the frequencies of the transmons such that energy exchange happens through capacitive coupling. More specifically, if we tune the frequency of the first qubit ω_{01}^{q0} to match the

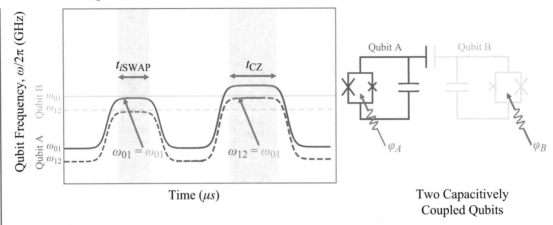

Figure 2.13: Two-qubit interactions for two capacitively coupled transmons. **Left:** Two-qubit gates are implemented with resonance of qubit frequencies. Shown here are how qubit frequencies are tuned for iSWAP gate and CZ gate. **Right:** Circuit diagram of two capacitively coupled transmon qubits.

frequency of the second qubit $\omega_{01}^{q}1$ (often referred to as tunning the two qubit *on resonance*), we have enabled a periodic population swap between the basis $|10\rangle$ and $|01\rangle$, as shown in Figure 2.13. This is sometimes called XY interaction:

$$XY[t] = e^{-i\frac{g}{2}(\sigma_x\sigma_x+\sigma_y\sigma_y)t} = \begin{pmatrix} 1 & 0 & 0 & 0 \\ 0 & \cos(gt) & -i\sin(gt) & 0 \\ 0 & -i\sin(gt) & \cos(gt) & 0 \\ 0 & 0 & 0 & 1 \end{pmatrix},$$

where g is the coupling strength of the capacitor. Note that when we tune the qubit on resonance for time π/g, we obtain the iSWAP gate:

$$XY[\pi/2g] = \text{iSWAP} \equiv \begin{pmatrix} 1 & 0 & 0 & 0 \\ 0 & 0 & -i & 0 \\ 0 & -i & 0 & 0 \\ 0 & 0 & 0 & 1 \end{pmatrix}.$$

The $\sqrt{i\text{SWAP}}$ gate, which is equivalent to $XY[\frac{\pi}{4g}]$, is sometimes useful as well.

Alternatively, if we tune the frequency of the first qubit ω_{01}^{q0} to match the *secondary* frequency of the second qubit ω_{12}^{q1}, we enable the periodic population swap between basis states $|11\rangle$ and $|02\rangle$. This is hardly desirable, if we leave the population at $|2\rangle$ which is beyond the computation subspace ($|0\rangle$ and $|1\rangle$); a phenomenon known as "leakage." However, when the population is swapped to $|02\rangle$ and back $|11\rangle$, we gain a $e^{-i\pi}$ phase on $|11\rangle$. In effect, we have accomplished the CZ gate. In fact, one can choose the interaction trajectory such that an arbitrary phase $e^{-i\theta}$ is gained on $|11\rangle$:

$$CZ_\theta \equiv \begin{pmatrix} 1 & 0 & 0 & 0 \\ 0 & 1 & 0 & 0 \\ 0 & 0 & 1 & 0 \\ 0 & 0 & 0 & e^{-i\theta} \end{pmatrix}.$$

Both the iSWAP and the CZ gates are useful primitives, as they can be used to implement the CNOT gate:

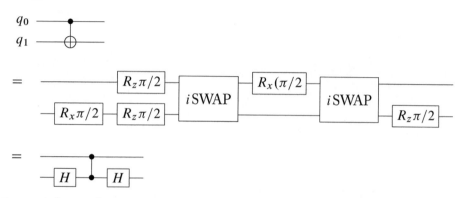

One can also implement two-qubit gates using only microwave control. Instead of tuning the qubit frequencies via external magnetic fluxes, the *CR (cross-resonance) gate* [129] achieves the two-qubit interaction on two (fixed-frequency) transmons coupled with a resonator via the qubit drive. In particular, if we drive the first qubit at the frequency of the second qubit, then the Rabi oscillation of the second qubit will have a frequency dependent on the state of the first qubit. This is sometimes referred to as the *ZX interaction*:

$$CR_\theta \equiv ZX[\theta] = e^{-i\frac{\theta}{2}\sigma_z \otimes \sigma_x} = \begin{pmatrix} \cos(\theta/2) & -i\sin(\theta/2) & 0 & 0 \\ -i\sin(\theta/2) & \cos(\theta/2) & 0 & 0 \\ 0 & 0 & \cos(\theta/2) & i\sin(\theta/2) \\ 0 & 0 & i\sin(\theta/2) & \cos(\theta/2) \end{pmatrix}.$$

The CR gate can be used to implement a CNOT gate (up to a phase $e^{i\pi/4}$):

2.4.3 OTHER PROMISING IMPLEMENTATIONS

Besides superconducting and trapped-ion architectures, there are other promising QC platforms. Due to limited space, we will not describe these platforms in detail, but refer the reader to the relevant literature: semiconductor spin qubits [130–133], linear optics [134–136], and Majorana qubits [137–140].

CHAPTER 3

Quantum Application Design

What kinds of computational problems are quantum computers good at solving? What are the major challenges for algorithm designers in the NISQ era? This chapter is devoted to updating the readers with recent progress and upcoming challenges in near-term algorithm design. Quantum machines will be useless, if we do not know what algorithms we can run on them. So a big part in the race to practical quantum computing is to develop algorithms that run on NISQ computers. We lead with a section illustrating the general features of quantum information processing, how to exploit the so-called quantum parallelism, and how to evaluate the cost of a quantum program. After answering the above questions, we introduce a few medium-scale quantum algorithms such as the Deutsch-Josza algorithm and the Bernstein–Vazirani algorithm, as well as some other classes of algorithms tailored for NISQ computers, such as the Variational Quantum Eigensolver (VQE) and the Quantum Approximate Optimization Algorithm (QAOA). We conclude this chapter with a survey on promising quantum applications. In the next 5–10 years, we probably won't have quantum processors built into our laptop or smart phone. But they will become useful for synthesizing better drugs and material, producing lower-energy fertilizer, solving optimization problems more efficiently, and so on.

3.1 GENERAL FEATURES

We start the discussion on quantum algorithms with a summary of their general features. The power of quantum computing can be viewed as ultimately coming from the ability to encode exponential computational space into just a linear number of computational units—the state of an entangled n-qubit quantum system is represented by 2^n complex coefficients. Each constant-time operation on the quantum system, in principle, manipulates non-trivially all 2^n complex coefficients, through $O(n)$ independent knobs (e.g., I, X, Y, and Z controls on each qubit). At the end of a quantum algorithm, one expects to measure the n qubits and obtain a random outcome of n classical bits. The art of designing quantum algorithms is thus to find transformations on the state of the qubits after which the final measurements yield the desired outcome with high probability. In the following, we elaborate on this design process, followed by a few remarkable example quantum algorithms.

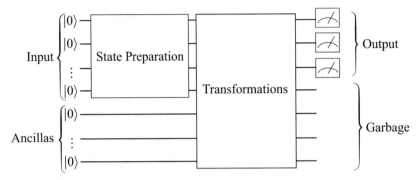

Figure 3.1: A typical quantum circuit implementation of a quantum algorithm.

3.1.1   THE COMPUTING PROCESS

The computational process of a quantum computer generally requires manipulating qubits in a particular manner, so as to fully exploit the power of quantum information processing. A prescription commonly observed in quantum algorithms is:

- efficiently encode information into a small number of qubits;

- cleverly build up entanglement and interference during the algorithm; and

- design a final measurement that yields desired outcomes with high probability.

As such, the implementation of a quantum algorithm can typically be expressed in terms of a quantum circuit in Figure 3.1.

Even though a quantum computer is said to have the power of manipulating an exponentially sized computational space, we must not start with an exponentially complex initial state. Otherwise the state preparation circuit that puts the qubits into such state would be extremely expensive.

In the following section, we describe in greater detail one of the key ingredients in quantum information processing: quantum parallelism. To better illustrate this phenomenon, we switch our mindset to the query model of computation.

3.1.2   THE QUERY MODEL AND QUANTUM PARALLELISM

A query model involves a black-box function (e.g., $f : \{0, 1\}^n \rightarrow \{0, 1\}^m$) called an *oracle* (as shown in Figure 3.2). An algorithm is trying to learn some properties of f by evaluating f on a set of inputs but not examining how f is implemented internally. We can imagine the oracle O_f is implemented as some private function (unknown to its user). The only way to learn the oracle is by passing some inputs to it and analyzing the outputs. The number of queries required by the algorithm is defined as the *query complexity* of the algorithm. In a nutshell, a query algorithm is a prescription of how to (efficiently) learn properties of a function f of interest.

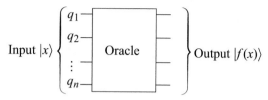

Figure 3.2: An oracle (or black-box function) that computes $f(x)$.

Here, we give some example problems well studied in the query model.

- *Searching Problem:* Find x such that $f(x) = 1$ [13].

- *Period-finding Problem:* Find the period of f when inputs x are ordered from 0...0 to 1...1, that is, find p such that $f(x) = f(x + p)$ for all x [12].

- *Collision Problem:* Find x, y such that $f(x) = f(y)$ (often used for analyzing graphs) [141, 142].

Quantum Query Algorithm vs. Classical Query Algorithm
Intuitively, the advantage of a quantum query algorithm is that we can pass several inputs as a superposition quantum state to the oracle (as shown in Figure 3.2) at one time. In return, we get a new superposition state as output. In contrast, a classical oracle only accepts a single input at each time, which limits the information we can get from each query.

Quantum Oracles
Consider a function $f : \{0, 1\}^n \to \{0, 1\}^m$, meaning that f takes an n-bit input and returns an m-bit output. In Chapter 2, we showed that f must be made reversible when implemented in a quantum circuit. In order to make the oracle reversible, we add some output qubits to the circuit to make the output width and input width identical. In the following quantum circuits, we call $|x\rangle$ the input qubits (or input registers) and $|y\rangle$ the output qubits (or output registers, or ancilla).

- *XOR oracle.* Oracle O_f is the XOR oracle that implements function f. It transforms a quantum state from $|x\rangle \otimes |y\rangle$ to $|x\rangle \otimes |y \oplus f(x)\rangle$. Note that if $|y\rangle = |0\rangle$, we obtain

$|y \oplus f(x)\rangle = |f(x)\rangle.$

- *Phase oracle.* $O_f^{\pm}$ is called the phase oracle of function f. It transforms a quantum state from $|x\rangle \otimes |y\rangle$ to $|x\rangle \otimes (-1)^{f(x) \cdot y} |y\rangle$, where $f(x) \cdot y = \sum_i f(x)_i y_i \mod 2$ is the inner product of the two bit-strings.

The XOR oracle and the phase oracle are equivalent, which means that each of them can be simulated by the other. We can build a phase oracle using the XOR oracle and vice versa.

It is sometimes convenient to simplify the oracle for $f : \{0, 1\}^n \to \{0, 1\}^m$ when $m = 1$, that is, f is a decision problem (1 for true and 0 for false). This is because the output is:

$$|x\rangle \otimes |- \oplus f(x)\rangle = \frac{1}{\sqrt{2}} |x\rangle \otimes (|0 \oplus f(x)\rangle - |1 \oplus f(x)\rangle).$$

Considering $f(x)$ is either 0 or 1, we have the output:

$$\frac{1}{\sqrt{2}} |x\rangle \otimes (|0 \oplus f(x)\rangle - |1 \oplus f(x)\rangle) = |x\rangle \otimes (-1)^{f(x)} (|0\rangle - |1\rangle) =$$

$$\frac{1}{\sqrt{2}}|x\rangle \otimes (-1)^{f(x)}(|0\rangle - |1\rangle) = |x\rangle \otimes (-1)^{f(x)}|-\rangle = (-1)^{f(x)}|x, -\rangle.$$

Hence, the oracle maps from $|x, -\rangle$ to $(-1)^{f(x)}|x, -\rangle$. As we don't care about the last qubit, we can simplify the system by ignoring the output qubits as follows:

Notice that a quantum oracle differs from a classical one by the ability to apply the function f to a superposition of states simultaneously, a phenomenon commonly referred to as "quantum parallelism,"

$$\sum_i |x_i\rangle \rightarrow \sum_i (-1)^{f(x_i)}|x_i\rangle.$$

3.1.3 COMPLEXITY, FIDELITY, AND BEYOND

In quantum computing, there are generally two notions of costs: (i) device-independent computational complexity and (ii) device-dependent implementation cost. The two notions typically rely on different assumptions, yet have close connection with each other, e.g., circuit complexity (i.e., number of gates) is related to running time (i.e., number of steps).

Computational Complexity
Following the definition of classical computational complexity, we can define quantum computational complexity, in which we characterize general properties of quantum algorithms. As we shall see in Chapter 6, we can efficiently simulate a quantum circuit using one universal gate set with another universal gate set, which allows us to define complexity classes *independent* of the implementation details, such as the choice of gate set, and the accuracy of the quantum gates.

The computational complexity of quantum algorithms is generally analyzed in two different styles, namely the time complexity and the query complexity.

- *Time complexity.* The time complexity of a unitary transformation U is related to the number of gates of the smallest circuit that implements U. In most cases it is hard to find the time complexity for a quantum algorithm, as we need to prove some circuit is an implementation of U and also there are no other smaller circuits for U.

- *Query complexity.* Query complexity is the number of times an algorithm needs to query a given black-box function (often called an oracle) to solve a problem. For many query-based quantum algorithm, such as Grover's algorithm, the query complexity is easier to analyze, as we introduced the query model in Section 3.1.2.

Before we introduce the quantum computational complexity class BQP (bounded-error quantum polynomial time), we first review two classical complexity classes, namely P and BPP, using the circuit model of computation.

Definition 3.1   The complexity class P is the class of all decision problems solvable by a polynomial-size uniform circuit family (with classical AND/OR/NOT gates) $\{C_n : n \in \mathbb{N}\}$ deterministically.

Here, decision problems are functions that take n-bit input and produce 1-bit output (i.e., an Yes/No answer). Polynomial-size uniform means there exists a polynomial-time deterministic Turing machine that outputs a description of the polynomial-size circuit on all inputs. This is the class of problems that are considered as *efficiently* solvable.

Definition 3.2   The complexity class BPP (bounded-error probabilistic polynomial time) is the class of all decision problems solvable by a polynomial-size uniform random circuit family (with classical AND/OR/NOT gates and coin flips) $\{C_n : n \in \mathbb{N}\}$ with high probability.

Bounded (two-sided) error means that the circuit C_n solves a Boolean function such that

- $\forall x \in \{0, 1\}^n$, if $f(x) = 1$, then $\mathbf{Pr}[C_n[x] = 1] \geq \frac{2}{3}$,
- $\forall x \in \{0, 1\}^n$, if $f(x) = 0$, then $\mathbf{Pr}[C_n[x] = 1] \leq \frac{1}{3}$.

Notice that, in contrast to P, BPP allows coin flips in its circuit, and is the notion of efficiently solvable in randomized computation.

Definition 3.3   The complexity class BQP (bounded-error quantum polynomial time) is the class of all decision problems solvable by a polynomial-size uniform quantum circuit family (with universal quantum gates) $\{C_n : n \in \mathbb{N}\}$ with high probability.

From BPP to BQP, we replace the random circuit with a quantum circuit. Notice that the class BPP is contained in BQP (i.e., BPP $\subseteq$ BQP), simply because we can simulate a fair coin flip by preparing a superposition state $\frac{1}{\sqrt{2}}(|0\rangle + |1\rangle)$ and then measure in the computational basis $\{|0\rangle, |1\rangle\}$. The relations of BQP to many other classical computational classes are still exciting open research problems. We provide another proof of such relation, BQP $\subseteq$ PSPACE (polynomial space) later in Chapter 9. Instead, we now turn our discussion to realistic implementation cost of quantum algorithms on a given hardware.

Implementation Cost

In practice, we care about the resource requirement to implement an algorithm on the physical device. There are several main concerns regarding implementations of quantum algorithms. The first issue we encounter is *precision*. Relaxing the idealized assumption that quantum gates are implemented with perfect accuracy, we must consider the impact of noise on the quantum circuit implementations. The accuracy of a quantum circuit implementation is typically defined by some distance measure between an ideal vs. a noisy process. The commonly used distance measures can be categorized in two classes: (i) input dependent and (ii) input independent. When the input state is given, we may quantify the accuracy of an implementation by measuring the distance in the output quantum states:

$$D\left(U_{noisy}\rho_{in}U_{noisy}^{\dagger}, U_{ideal}\rho_{in}U_{ideal}^{\dagger}\right),$$

where $D(\cdot)$ is a distance metric. Some common metrics between two quantum states are *fidelity* and *trace distance*, which are defined in Chapter 8. Alternatively, we can also directly measure the distance between two processes, regardless of the input state:

$$D\left(U_{noisy}, U_{ideal}\right).$$

It is also reasonable to compare measurement outcomes of two quantum circuits, instead of the quantum states themselves; this is typically the case in the context of classical simulation of quantum computation. Since measurement outcomes are essentially probability distributions, we adapt metrics such as *total variation distance* for comparing two distributions, as defined in Chapter 9.

When all of the above metrics are difficult calculate, for example, when a quantum computer has insufficient power (e.g., not enough qubits for the target quantum circuit) or when an efficient classical simulator is unavailable, a common last resort is to estimate the *success rate* of a quantum circuit using efficiently computable noise models. For instance, one can estimate the (worst-case) success rate of a quantum circuit under qubit decoherence and gate noise:

$$P_{success} = \Pi_{g \in G}(1 - \epsilon_g) \cdot \Pi_{q \in Q}(1 - \epsilon_q),$$

where ϵ_g is the average gate error rate, and ϵ_q is captured by modeling T_1 and T_2 during idle or gate time, as shown in Chapter 2.3.3.

The second issue in implementing quantum algorithms is *resource cost*. In practice, we can further optimize a circuit implementation by strategically choosing the most robust qubits and most robust gates, or by shortening the running time of the circuit. As such, some resource metrics for describing the cost of an implementation are motivated. We list some of them as follows.

- *Qubit count.* Also referred to as circuit width. It represents the number of qubits the algorithm needs, and limits the dimension under which the algorithm operates.

- *Gate count.* The number of quantum gates used in the given algorithm. For NISQ algorithms, this is a useful metric for obtaining a general sense of their success rate, as gate errors are the most dominant source of noise in today's NISQ machines.

- *Circuit depth.* The number of time steps the algorithm uses. Note that we need to find the deepest path in the circuit from the input to the output. This is often referred as "critical path" generalized from classical Boolean circuits. Circuit depth is closely related to the "running time" of a quantum circuit. Higher circuit depth typically correlates with lower success rate, because deeper circuits suffer from noise due to the higher gate count as well as due to the higher likelihood of experiencing qubit decoherence.

- *Communication cost.* Because of physical limits, we need to take extra effort to communicate the qubits to enable the actual computation. This often happens when the physical qubits are not fully directly connected in a quantum computer. For example, if in a super conducting machine, two qubits are not close to each other and there is no physical connection between them, we need to swap one of them closer to another to apply later computation.

- *Spacetime volume.* At a high level, it represents "*space × time*" of an algorithm. We can also define the quantum volume for a quantum computing machine as *#qubits × depth*. Quantum volume often serves as a quantity with which we can compare the performance of two different machines. Note that the error rate and the topology also contributes to the quantum volume.

3.2   GATE-BASED QUANTUM ALGORITHMS

To many, the most intriguing discovery in quantum computing is the fact that quantum algorithms can solve certain problems that are inefficient or infeasible on a classical computer. In this section, we highlight a selection of interesting problems to provide the reader with a general sense of the key ingredients in quantum algorithms. For example, the *Deutsch–Josza algorithm* [143] is one of the first quantum algorithms with shown *quantum advantage*, that is a gap exists between the computational complexity of best-known classical algorithm and that of the quantum one. Another example we introduce in this section is the *Bernstein–Vazirani algorithm* [9]. Both algorithms show impressive speedup over the best-known classical algorithms, but there are several caveats. First, neither problem has known applications, as they solve very specific, rather contrived mathematical problems. Second, the comparisons were done using the query model which is not exactly how classical functions are usually evaluated, and not directly related to the familiar *running time* concept. Third, our analyses are based on comparing quantum algorithms with classical deterministic algorithms. Many examples here can be made more efficient with randomized algorithms. But it is generally believed that for any problem p, its quantum query complexity is smaller than (or equal to) its classical randomized complexity, which is smaller than (or equal to) its classical deterministic complexity. Nonetheless, it remains an open problem to determine the relation between the three models of computation.

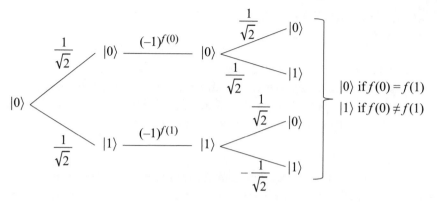

Figure 3.3: Feynman path diagram of Deutsch's algorithm.

3.2.1   DEUTSCH–JOSZA ALGORITHM

Let us begin with a simple quantum algorithm called "Deutsch's algorithm," and then move on to its generalized version.

Deutsch's Algorithm [6]

Problem statement: Given two unknown bits b_0, b_1 and an oracle that implements function f s.t. $f(0) = b_0$, $f(1) = b_1$, we want to determine the "parity" of b_0 and b_1, i.e., whether b_0 and b_1 are different or the same.

Classical solution: We can use *two* queries: $f(0)$ and $f(1)$, and then compare the results to determine the parity of b_0, b_1.

Quantum solution: We use the following quantum circuit to solve the problem (in the simplified phase oracle model):

$$|0\rangle \ -\boxed{H}-\boxed{O_f^{\pm}}-\boxed{H}-\boxed{\measuredangle}$$

Let us analyze the circuit with the *Feynman Path Diagram* depicted in Figure 3.3.

The quantum states for each step in the circuit are listed below:

1. $|0\rangle$

2. $|+\rangle = \frac{1}{\sqrt{2}}(|0\rangle + |1\rangle)$

3. $\frac{1}{\sqrt{2}}\left((-1)^{f(0)}|0\rangle + (-1)^{f(1)}|1\rangle\right) = \frac{(-1)^{f(0)}}{\sqrt{2}}\left(|0\rangle + (-1)^{f(1)-f(0)}|1\rangle\right)$

$$= \begin{cases} (-1)^{f(0)}|+\rangle & \text{if } f(1) = f(0) \\ (-1)^{f(0)}|-\rangle & \text{if } f(1) \neq f(0) \end{cases}$$

4. $\begin{cases} (-1)^{f(0)}|0\rangle & \text{if } f(1) = f(0) \\ (-1)^{f(0)}|1\rangle & \text{if } f(1) \neq f(0) \end{cases}$

The constant factor $(-1)^{f(0)}$ is called the global phase of a qubit, which does not matter when measured. We can see that the result of the final measurement will always be 0 if $b_0 = b_1$ and will always be 1 otherwise. Thus, the problem is solved within *one* query of the oracle, as opposed to two queries in the classical case.

Deutsch–Josza Algorithm [143]

The constant speedup as shown in Deutsch's algorithm may not seem very impressive. However, the following generalization is certainly eye-catching, for it has a striking exponential speedup over any classical deterministic solution.

Problem statement: We have an oracle implementing a function $f : \{0, 1\}^n \to \{0, 1\}$ which is promised to be either constant ($f(x)$ is always 0 or $f(x)$ is always 1) or balanced (half of the inputs x, $f(x) = 0$, and the other half $f(x) = 1$). We want to determine whether it is constant or balanced.

Classical solution: In the classical world, we need to query $2^{n-1} + 1$ times (in the worst scenario), covering inputs for more than half of the domain $\{0, 1\}^n$ in order to determine whether f is balanced or constant. Should we use a randomized algorithm, we can achieve a bounded ϵ error on the result in $\log\left(\frac{1}{\epsilon}\right)$ queries.

Quantum solution [143]: We use a similar circuit to the one for Deutsch's algorithm:

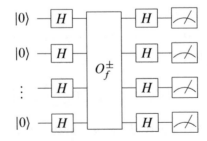

Again, we can track the quantum state with a Feynman path diagram (see Figure 3.4).

We write down the quantum states after each step of the algorithm as follows:

1. $|0\rangle^{\otimes n}$

2. $|+\rangle^{\otimes n} = \frac{|0\rangle+|1\rangle}{\sqrt{2}} \cdots \frac{|0\rangle+|1\rangle}{\sqrt{2}} = \frac{1}{\sqrt{2^n}} \sum_{x\in\{0,1\}^n} |x\rangle$

3. $\frac{1}{\sqrt{2^n}} \sum_{x\in\{0,1\}^n} (-1)^{f(x)}|x\rangle$

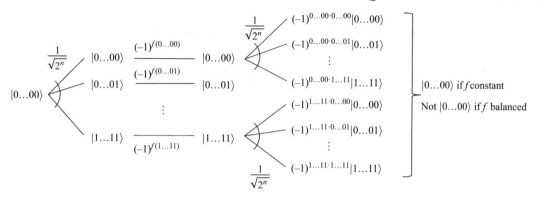

Figure 3.4: Feynman path diagram for Deutsch–Josza algorithm.

$$= \begin{cases} (-1)^{f(x)} \frac{1}{\sqrt{2^n}} \sum_{x \in \{0,1\}^n} |x\rangle = (-1)^{f(x)} |+\rangle^{\otimes n} & \text{if } f(x) \text{ constant} \\ \frac{1}{\sqrt{2^n}} \sum_{x \in \{0,1\}^n} (-1)^{f(x)} |x\rangle & \text{if } f(x) \text{ balanced} \end{cases}$$

$$4. \begin{cases} (-1)^{f(x)} |0\rangle^{\otimes n} & \text{if } f(x) \text{ constant} \\ \frac{1}{\sqrt{2^n}} \sum_{x \in \{0,1\}^n} (-1)^{f(x)} H^{\otimes n} |x\rangle & \text{if } f(x) \text{ balanced} \end{cases}$$

$$= \begin{cases} (-1)^{f(x)} |0\rangle^{\otimes n} & \text{if } f(x) \text{ constant} \\ \frac{1}{\sqrt{2^n}} \sum_{x \in \{0,1\}^n} (-1)^{f(x)} \frac{1}{\sqrt{2^n}} \sum_{y \in \{0,1\}^n} (-1)^{xy} |y\rangle & \text{if } f(x) \text{ balanced} \end{cases}$$

$$= \begin{cases} (-1)^{f(x)} |0\rangle^{\otimes n} & \text{if } f(x) \text{ constant} \\ \frac{1}{2^n} \sum_{x \in \{0,1\}^n} \sum_{y \in \{0,1\}^n} (-1)^{xy+f(x)} |y\rangle & \text{if } f(x) \text{ balanced} \end{cases}$$

We only focus on the amplitude of the quantum state $|0...0\rangle$. We can see that if f is constant, the amplitude is 1 (up to global phase), and if f is balanced, the amplitude is 0 (as $\sum_{x \in \{0,1\}^n} (-1)^{f(x)}$ cancels out). This means that after we measure the output, if the result is $0...0$, f must be constant, and if the result is something other than $0...0$, f must be balanced. The algorithm solves the problem with only one query, which means we have an exponential speedup over classical algorithm ($2^{n-1} + 1$).

3.2.2 BERNSTEIN–VAZIRANI ALGORITHM

In (the non-recursive) Bernstein–Vazirani Algorithm [9], we will see that there is no exponential speedup compared to the classical algorithm but a polynomial one.

Problem statement: Given an oracle access to $f : \{0, 1\}^n \to \{0, 1\}$ and a promise that the function $f(x) = s \cdot x = \sum_{i=1}^{n} s_i \cdot x_i \mod 2$, where s is a secret string that the algorithm is trying to learn.

Classical solution: Classically, we will have to try it out the brute-force way by giving n inputs, i.e.,

$$f(100...0) = s_1$$
$$f(010...0) = s_2$$
$$\vdots$$
$$f(000...1) = s_n$$

Quantum solution: In quantum computation, we can do this in just *one* query as shown in the circuit below:

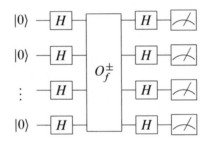

Notice that the circuit is identical to the one used in Deutsch–Josza algorithm; the only difference is in the oracle and the interpretation of the measurement results.

The states at different levels are:

1. $|0\rangle^{\otimes n}$

2. $|+\rangle^{\otimes n} = \frac{1}{\sqrt{2}} \sum\limits_{x \in \{0,1\}^n} |x\rangle$

3. $\frac{1}{\sqrt{2^n}} \sum\limits_{x} (-1)^{f(x)} |x\rangle = \frac{1}{\sqrt{2^n}} \sum\limits_{x} (-1)^{s \cdot x} |x\rangle$
   $= \frac{1}{2}(|0\rangle + (-1)^{s_1} |1\rangle) \otimes \frac{1}{2}(|0\rangle + (-1)^{s_2} |1\rangle) \otimes \cdots \otimes \frac{1}{2}(|0\rangle + (-1)^{s_n} |1\rangle)$.
   The state of the ith qubit thus depends on s_i: if $s_i = 0$ (or 1) then qubit i is $|+\rangle$ (or $|-\rangle$).

4. 0 if $s_i = 0$ and 1 if $s_i = 1$.

So far, we assume that such an oracle as used in the Bernstein–Vazirani algorithm exists without worrying about its practical implementations. Details about implementing oracles can be found in Chapter 6, Section 6.1.2.

3.3 NISQ QUANTUM ALGORITHMS

A class of quantum algorithms that are believed to be useful in the NISQ era, where technology limits the quantity and quality of qubits. Resources in a NISQ machine are scarce, not enough for applying existing QEC techniques. In the following, we show how NISQ algorithms boost their resilience to noises. In particular, we discuss two leading NISQ algorithms:

1. Variational Quantum Eigensolver (VQE) [46, 144, 145] and

2. Quantum Approximate Optimization Algorithm (QAOA) [68]

3.3.1 VARIATIONAL QUANTUM EIGENSOLVER (VQE)

The key to understand the VQE algorithm is the *variational principle*, which says that for any given vector $|\psi\rangle$,

$$\langle\psi|\, H\, |\psi\rangle \geq E_0,$$

H is hermitian, which implies that H has a complete set of eigenvectors which are orthonormal to each other. Let $|E_0\rangle, |E_1\rangle, .. |E_n\rangle$ be the eigenvectors of H with eigenvalues $E_0, E_1, .. E_n$, respectively, with E_0 being the smallest. Then any given vector can be written as a superposition of these eigenvectors, i.e.,

$$|\psi\rangle = C_0\,|E_0\rangle + C_1\,|E_1\rangle + ... C_n\,|E_n\rangle.$$

So for any given $|\psi\rangle$, we have

$$
\begin{aligned}
\langle\psi|\, H\, |\psi\rangle &= (C_0\,\langle E_0| + C_1\,\langle E_1| + ... C_n\,\langle E_n|)H(C_0\,|E_0\rangle + C_1\,|E_1\rangle + ... C_n\,|E_n\rangle) \\
&= C_0^2 E_0 + C_1^2 E_1 + ... C_n^2 E_n \\
&\geq E_0.
\end{aligned}
$$

As shown in Figure 3.5, to find the lowest eigenvalue, we keep checking E for different values of $|\psi\rangle$ till we get the minimum. The variational principle means that any guess would get us closer to the true lowest value. In VQE, we parametrize ψ using θ and find the value for θ which gives the smallest value for H which is similar to classical optimization. This can be summarized as guess and check, where the check is measuring the $\langle H\rangle$ and the guess is using a classical optimizer to select the next value of θ based on previous results. Here, $|\psi(\theta)\rangle$ is the ansatz and we have two ways of getting the ansatz: (i) hardware/machine ansatz and (ii) problem/physics ansatz.

One good analogy is learning Ping Pong. The above ansatz can be understood as different approaches that one can take to learn it. One strategy would be to go to a coach and learn from the first principles which is similar to "Problem/physical ansatz." The second approach is building up from what you already know, for example tennis. Both has its pros and cons and the idea is to select the ansatz based on the problem. If our choice of ansatz doesn't matter much, then we can go for the one which has the best machine performance. But if the choice of the ansatz matters then we have to design the circuit accordingly.

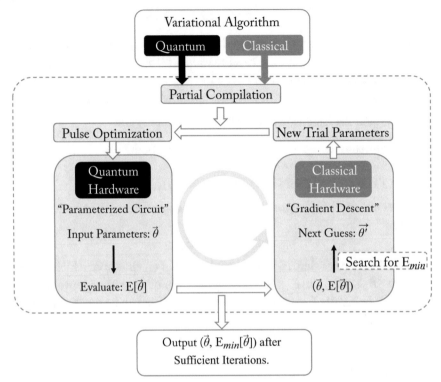

Figure 3.5: Illustration of a variational quantum algorithm that alternates between a quantum circuit and a classical optimizer. In this process, the quantum circuit (parameterized by $\vec{\theta}$) evaluates some cost function $E[\vec{\theta}]$, and the classical optimizer gradient descends for the next set of parameters.

3.3.2   QUANTUM APPROXIMATE OPTIMIZATION ALGORITHM (QAOA)

Both of the algorithms solve the same problem which is finding the lowest Eigenvalue of a hermitian matrix H. In the context of quantum simulations, such matrix H is called the Hamiltonian of a quantum system. The Hamiltonian is an operator which tells us about the energy of the system at time t. Since H is a hermitian matrix, it has real eigenvalues and orthonormal eigen vectors:

$$H\,|E_n\rangle = E_n\,|E_n\rangle\,.$$

Out of all these eigenvalues, the one with lowest energy is called ground state energy. The challenge is to find this ground state of the $2^n \times 2^n$ matrix Hamiltonian.

MaxCut/Clustering is one example where we can use QAOA [68] to find the solution. In this problem we are required to maximize the weight of the edges crossing the cuts. A cut

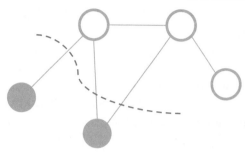

Figure 3.6: A MaxCut of a graph, where edges are cut if they connect a gray node with a white node.

is made on an edge whenever it is connected to vertices of two different colors. This can be rewritten as as an eigenvalue problem with a Hamiltonian given by

$$H = \frac{1}{2} \sum_{\text{edges } i,j} \left(I - Z_i Z_j \right).$$

Here, Z_i and Z_j are the Pauli Z matrix for the i th and j th vertex. The maximum cut then would be same as the ground state eigenvalue of this Hamiltonian. For example, we can consider just 2 edges connected by an edge. Then Hamiltonian is given by

$$H = \frac{1}{2} \left(\begin{bmatrix} 1 & 0 & 0 & 0 \\ 0 & 1 & 0 & 0 \\ 0 & 0 & 1 & 0 \\ 0 & 0 & 0 & 1 \end{bmatrix} - \begin{bmatrix} 1 & 0 \\ 0 & -1 \end{bmatrix} \otimes \begin{bmatrix} 1 & 0 \\ 0 & -1 \end{bmatrix} \right) = \begin{bmatrix} 0 & 0 & 0 & 0 \\ 0 & -1 & 0 & 0 \\ 0 & 0 & -1 & 0 \\ 0 & 0 & 0 & 0 \end{bmatrix}.$$

This matrix has eigenvectors $|00\rangle$ (no crossing) with eigenvalue 0 and $|01\rangle$ (with crossing) with eigenvalue -1. Here 0 corresponds to edges connecting the same color and -1 represents the one which connects different colors. Figure 3.6 is an example MaxCut of a graph of five nodes.

3.4   SUMMARY AND OUTLOOK

Quantum computing opened an entirely new way of solving computational problems efficiently. It is so far the only model that violates the extended Church-Turing thesis—remarkably a quantum computer can solve some computational tasks in exponentially fewer steps than the best classical computer. Large gate-based algorithms such as Shor's algorithm [11, 12] and Grover's algorithm [13] have shown practical potential of quantum computers. These algorithms have also generated public concerns in computer and network security—current public key cryptosystems rely on the hardness of factoring large numbers and computing discrete logarithms, which can be solved exponentially faster on an idealized quantum computer.

In the near-term, quantum devices are still limited in fidelity and size. These NISQ devices [14] are not fault tolerant and thus cannot implement the algorithms developed for ideal quantum computers.

Due to the resource limitations of NISQ machines, benchmarking is typically performed on small gate-based algorithms (such as the Deutsch–Josza algorithm and the Bernstein–Vazirani algorithm), variational quantum eigensolvers (applied to various electronic systems), quantum approximate optimization algorithms (applied to small problem instances), and small random circuits. One of the biggest challenges of the NISQ era is to develop algorithms that run on NISQ computers.

Remarkably, the competitions between quantum and classical algorithms have formed a virtuous cycle. A recent example is the invention of the HHL algorithm for solving linear systems [146] thought to have exponential speedup over any classical algorithms. Further, a quantum-inspired classical algorithm [147] was discovered with significant improvement over existing classical algorithms that matches its quantum counterpart in some cases. Despite the loss of quantum advantage in those cases, the consideration of a quantum solution has advanced our understanding of the nature of this problem resulting in a classical improvement.

Further Reading

The focus of this book is on near-term applications. A promising approach to reduce the cost of quantum algorithms is to use an approximate or heuristic approach to solve problems, giving rise to the hybrid classical-quantum algorithms, such as variational algorithms for simulating molecules and materials [46, 144, 145, 148], optimization algorithms [68], and machine learning [149–151].

A near-term milestone for quantum computing is a demonstration of quantum advantage, sometimes referred to as quantum supremacy, where experimentalists want to show a quantum computer can solve some problem exponentially faster than the best classical computer. To demonstrate this, one would need not only to build a quantum machine powerful enough to run experiments, but also to choose a test problem that is simple enough for a quantum computer but hard enough for a classical one. Aaronson and Arkhipov proposed in 2010 the Boson Sampling algorithm [152]. A small experiment with six photons has been implemented [153], but demonstration at a larger scale remains challenging [154]. Google proposed to use random circuit sampling to demonstrate quantum supremacy [155], whose classical hardness was later made rigorous by Bouland et al. [156].

PART II

Quantum Computer Systems

CHAPTER 4

Optimizing Quantum
Systems–An Overview

The second part of this book is dedicated to techniques for optimizing quantum computing at a systems level. In this chapter, we describe the layers of a quantum computer system. In subsequent chapters, we give examples of key optimizations that work across these layers, strategically trading software complexity for compute efficiency.

Developments in the theory of quantum algorithms and the implementation of quantum hardware in the past few years have been truly remarkable. But there are still formidable challenges lying ahead. So far, an enormous gap exists between the resources required by many discovered algorithms, and the resources available in today's devices. We will have to learn to execute large quantum algorithms under highly-constrained conditions. It is of paramount importance that we optimize for the resource consumption and success rate of a quantum program via sharing of information throughout the software-hardware stack. Such information includes characteristics of the target application and the underlying hardware, for example. An overarching theme in quantum computer systems research in the NISQ era will be *software-hardware co-design*, or in other words, vertical integration across the systems layers. A family of techniques across many layers will be needed. Each and every optimization will play a vital role in enabling practical quantum computing. Indeed, this is the emphasis of this book.

4.1   STRUCTURE OF QUANTUM COMPUTER SYSTEMS

The aim of this section is to provide a bird's-eye view of the key components in quantum computer systems. After introducing the architecture layers of a quantum computer, we shed light on where research opportunities lie in the NISQ era for computer system researchers and engineers. Quantum computing is at a similar stage of development as classical computing in the 1950s. Today's QC systems consist roughly of three essential components, namely the three layers in quantum computer architecture: application layer, systems software layer, and hardware layer, as shown in Figure 4.1. Today's classical computer systems manage highly complex hardware and software through layering abstractions. Going up through the systems stack, each layer hides some implementation details and expose a manageable set of controls for the next layer.

In contrast, the development of quantum computer systems is still at its nascent stage. This is great for researchers, because there are so many interesting problems to be solved. It also means that resources are very scarce and that we are motivated to *break abstractions* and

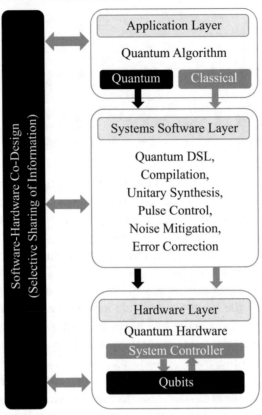

Figure 4.1: Selective sharing of information allows algorithms to use limited resource in NISQ hardware most efficiently.

pay for efficiency with greater software complexity. Even classical computing is backsliding a bit toward less abstraction as the end of Dennard scaling puts pressure on architectures to become more efficient. How much of what we learn in the next fivfive years will carry forward to a future of much larger quantum machines? Perhaps more than we might think, as it would be hard to imagine a future in which qubits and quantum operations are not costly. A functional quantum computer requires painstaking attention to the isolation and control over many qubits. Some physical details may always be exposed. The experience and lessons we learn about how to manipulate qubits in NISQ computers, be it at the algorithmic, systems, or hardware level, will pave the way for larger fault-tolerant quantum devices in the future. Noise resilience is not only for experimentalists who build the hardware to worry about; opportunities are ubiquitous in the entire systems stack. In fact, it is crucial for algorithm designers, systems architects, and software developers to take responsibilities in tackling this challenge together. It is expected that, in the NISQ era, a QC toolchain must break the traditional abstraction layers and use aggressive optimizations throughout the full systems stack. The key to successful execution of quantum algorithms on NISQ devices is to selectively share information across layers of the stack (from device specifics to application characteristics) such that programs can use the limited qubits most efficiently.

4.2   QUANTUM-CLASSICAL CO-PROCESSING

An important variation of quantum computing systems is their use as specialized hardware accelerators within a classical computation. Indeed, this hybrid co-processing approach will likely be the dominant structure of quantum systems for the foreseeable future.

While quantum computers are (currently) small and unreliable, a great way to exploit their special, but limited, abilities is to adopt a hybrid model [145] which leverages both quantum and classical computation. Almost all useful algorithms require some amount of classical pre-processing or post-processing. For example, Shor's algorithm has a series of classical arithmetic operations before and after the quantum order-finding subroutine. But perhaps the most promising example is in quantum chemistry, where Variational Quantum Eigensolver (VQE) algorithms perform a kind of heuristic search by iterating between a quantum machine and a classical supercomputer. The goal is to find the lowest energy state of a chemical compound (the ground state). As shown in Figure 3.5 of Chapter 3, we start from the best-known configuration of electrons from a classical computer and estimate the energy of that configuration using the quantum machine. This estimate is then given back to a classical computer to guide its search toward a configuration with lower energy. In this way, the quantum machine acts as an accelerator for the energy modeling part of the computation. By solving for lowest energy under different configurations and constraints, we can explore a range of molecular reactions.

This hybrid example has some great advantages. First, it sidesteps the "innovator's dilemma" by leveraging an initial guess derived from our incumbent classical technology, rather than directly competing with that technology. Second, hybrid algorithms break a long program into multiple iterations of short programs, which allows us to effectively utilize the limited number of instructions a quantum machine can reliably execute. Third, it allows us to pick small but classically challenging problems (chemical compounds) that can be represented in a small number of quantum bits. In order to determine which orbitals the electrons are in, Nature only uses n electrons to "model" n electrons, whereas classical computers require combinatorially k^n bits to model n electrons, but quantum computers only need kn qubits to model n electrons. Fourth, we have a clear measure of success, as we know that classically-computed ground state energy can be significantly higher than experimentally-observed values, even for small compounds. If our hybrid approach can get closer to experimental values, then our quantum machine has helped compute something not computable classically! Long-term, improved understanding of molecular reactions could lead to better materials, more efficient photovoltaics, and lower-energy fertilizer production.

Even as quantum machines scale, quantum algorithms are likely to be specialized, making the quantum device a very domain-specific accelerator. Most practical applications will still require a combination of general classical and specialized quantum processing to be useful.

The hybrid approach implies a number of interesting research challenges for system designers. Traditional quantum algorithms can be statically compiled with a high level of optimization using known input parameters. With hybrid algorithms, some of a quantum program's input parameters can change each iteration. For example, a compiler may spend hours optimizing for quantum instructions that include quantum rotations for specific input angles to solve a chemistry problem, but now we find that the angles change every iteration. This suggests that we need a partial compilation strategy in which programs are optimized for unchanging parameters, but then quickly re-optimized each iteration for parameters that change.

Hybrid algorithms also require more thought to be given to hardware and software communication mechanisms between quantum and classical hardware, as well as how such ensembles might be presented as compute resources to users. IBM was the first to make a physical quantum machine accessible via the cloud, which has greatly grown the quantum computing community and allowed research into how to adapt to the physical properties of real machines. The IBM machines, however, are cumbersome for hybrid computation, as the batch queue interface is really designed for stand-alone quantum programs and the latency to couple with classical computation is long.

4.3   QUANTUM COMPILING

A quantum compiler aims to efficiently express a high-level quantum program using instructions that a quantum machine recognizes and natively supports, balancing practical *architectural constraints*.

A quantum algorithm is implemented in a *quantum domain-specific language* (QDSL). The quantum compiler translates the high-level program into *quantum assembly code* (QASM) that can be executed on a target hardware. This is accomplished through a series of transformations and optimizations on a *quantum intermediate representation* (QIR) of a program. Finally, at the lowest level, machine-level instructions that orchestrate the hardware control pulses are scheduled and optimized.

For a program to be realizable on a given hardware, a number of architectural constraints must be satisfied. This typically means considering the following practical aspects.

- *Instruction set.* There are certain limited number of quantum instructions that are supported in a given architecture. A compiler should aim to translate high-level quantum programs using the supported instruction set. In most cases, this instruction set is "Clifford+T" gates, comprised of the CNOT (controlled-NOT) gate, X (NOT) gate, H (Hadamard) gate, and T ($\pi/8$-phase) gate. This is a common set for most gate-based NISQ machines, as well as large-scale FT machines (e.g., with surface code error correction). Some NISQ compilers choose to target directly the physical analog pulses for improved hardware control. The detailed discussions on pulse control have been deferred to Chapter 7.

- *Qubit communication.* A quantum algorithm is hardly interesting if it can be implemented with only single-qubit gates, as two-qubit gates (or multi-qubit gates) provides the entangling power between qubits. Two-qubit gates are implemented by qubit-qubit interaction/communication. Qubit communication has different meanings in the NISQ vs. the FT contexts—usually for a NISQ machine, not all qubits can directly interact with each other, two qubits interact by moving closer to one another via a chain of swap gates until they are directly connected hence allowed to interact. The time to complete a swap chain is proportional to the length of the chain. In FT machines, qubit interactions are accomplished through fault-tolerant operations depending on the error correcting codes (such as braiding and lattice surgery in surface-code error-corrected devices[1]). With today's technology, building large-scale quantum machines with all-to-all qubit connectivity is shown to be extremely challenging. The latest effort from IonQ [56] offers a machine with 11 fully connected qubits using trapped-ion technology. Superconducting machines, for instance by IBM [51] and Rigetti [158], typically have *much* lower connectivity. Any scalable proposal would involve an architecture of limited qubit connectivity and a model for resolving long-distance interactions, hence inducing communication costs. This constraint is sometimes referred to as "device topology."

- *Hardware noise.* Another important consideration for compiling quantum programs is to minimize errors caused by hardware noise. Errors under consideration typically include *memory errors* (caused by decoherence of qubits) and *gate errors* (caused by imprecise control of gates). In general, the longer the program runs, the higher the chance that the qubits experience decoherence. The more gates are applied, the lower the chance that the program succeeds at the end. In today's technology, a two-qubit gate proves be challenging, hence it is one of the dominant sources of error. A compiler normally aims to express a quantum program in fewer qubits, or fewer number of gates, or shorter circuit depth, etc. Note that these targets are non-exclusive, sometimes conflicting, in which case the compiler would need to balance between the constraints. More advanced noise-aware compilers have also been proposed. For example, in NISQ machines, some qubits are more robust then the others, so picking the longer-lived qubits to perform important computation can improve the overall success rate.

- *Available parallel control.* Depending on the technology that implements the qubits, a compiler can be constrained by the available parallelism. The parallelism limitation is

---

[1]Braiding and lattice surgery are techniques that implement gates between logical qubits on the surface code lattice. Details are omitted as they are out of the scope of the book; we refer the interested reader to Chapter 8 and other tutorials [22, 27, 157] for basics of quantum error correction and topological quantum codes.

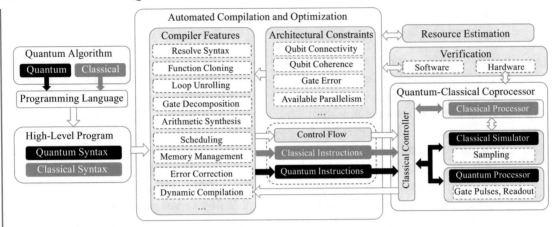

Figure 4.2: A detailed quantum compilation flow outlining the transformations and optimizations involved in a generic compiler.

usually the consequence of hardware control mechanism, or error mitigation protocols. For instance, the width of the tunable laser beams in a trapped-ion NISQ machine limits the number of independently controlable qubits, and thus the number of parallel single-qubit gates. Some error mitigation protocols dictate that no parallel gates are allowed when they are physically located close to each other, reducing crosstalk errors between them.

Figure 4.2 illustrates a typical quantum compilation toolflow. At its core, the quantum compiler passes a high-level quantum program through a series of optimizations, generating the most efficient and robust low-level executable (i.e., sequence of classical and quantum instructions) for the target hardware, balancing different architectural constraints. For historic reasons, more recent work typically targets NISQ-era architectures, but older work targets large FT architectures. Nonetheless, most techniques we introduce here generally apply to the different architectures. As hinted in Chapter 2, compilation for quantum machines is very similar to classical circuit synthesis. In the classical setting, we take some high-level language (C-like, Verilog, etc.), and compile it all the way down to instructions for transistors. This similarity is not pure coincidence; after all, the quantum circuit model of computation is generalized from the Boolean circuit model.

4.4   NISQ VS. FT MACHINES

Quantum compiling in the context of NISQ and FT era can be drastically different. This section aims to name a few examples of such differences, so that one does not confuse a technique for NISQ machines with another for FT machines, and vice versa.

Notably, quantum compiling in the NISQ era tends to be more *dynamic*. The emerging NISQ applications, such as the variational eigensolver and the quantum approximate optimization algorithm, have hybrid/interleaved classical and quantum processing—quantum circuits are parameterized with the parameters optimized by a classical algorithm. So the traditional model of compiling static quantum programs once would not work well in the NISQ context. [159] aims to save some compiling cost by reusing the partial synthesis results across the iterations of the algorithm.

Another difference is in the topology of the architecture and the model for resolving two-qubit interactions. As a result, communication costs will differ. In the context of a NISQ machine, the most frequently used approach to resolve a long-distance two-qubit gate is to move one qubit closer to the other through a chain of swaps. A SWAP gate can easily be implemented by three CNOT gates. In an FT machine, such as with the surface code error corrected architecture, we can resolve long-distance interactions between logical qubits through a process called braiding (i.e., movement and transformation of qubits) [160, 161]. Braiding has very different cost models than those of swapping. For instance, braids can extend to arbitrary length and shape in constant time, given that they never cross other braids; latency (i.e., time cost) of a swap chain is proportional to the length of the chain.

A third difference highlighted here is the choice of instruction set. Quantum circuit synthesis has been largely done in the context of Clifford+T gate set, due to its nice algebraic structures. Although that is a reasonable choice for FT machines (as Clifford gates are straightforward to implement fault-tolerantly for stabilizer error correction codes), it is not the ideal choice for NISQ machines. For example, NISQ machines can typically perform single-qubit rotations along one of the principal axes (e.g., z-rotations) to very high precision, while suffer on two-qubit gates such as CNOT gates. It remains an open problem in discovering optimal device- or application-adapted synthesis algorithms.

Last but not least, quantum compiling in the presence of noise has been under-studied. Integrating noise-awareness in circuit synthesis, gate scheduling, qubit mapping, pulse synthesis, and compiler validation are among the first challenges in quantum computer systems.

CHAPTER 5

Quantum Programming Languages

The concepts of data and operations for quantum computers could be drastically different from those of classical computers. For instance, the superposition principle dictates that quantum data (or qubit states) are intrinsically probabilistic as the information stored in a qubit can only be partially read out through an irreversible process called measurement which yields a probabilistic outcome. Operations on quantum data stored in one part of the memory could affect the data in another remote part due to a property called entanglement. What is perhaps more surprising is that quantum data cannot be duplicated into two independent copies, known as the no-cloning theorem. On top of those, one more layer of complexity is added by the fragility of quantum states. QC systems are susceptible to decoherence (i.e., spontaneous loss of quantum information in qubits) and operational errors. These are just some examples of the unique properties presented in quantum programs. They influence how a quantum program needs to be executed. Quantum algorithms typically involve a hybrid of classical and quantum processing. As such, a common strategy in quantum programming language design is to adapt and augment conventional programming language semantics and type systems to express the new properties of quantum programs.

5.1 LOW-LEVEL MACHINE LANGUAGES

At a lower level, a quantum hardware is controlled by instructions signaled by a classical host processor. The quantum assembly language (QASM) is a direct translation from a quantum circuit to a sequential description of quantum instructions for executing a quantum program. Although some existing low-level quantum languages are developed primarily with device-independent and software portability in mind, more and more attention is paid to exposing device specifics, such as the hardware native gates, device connectivity, and noise models, to the language itself and to its software toolchain.

One of earliest low-level quantum languages is called QASM [162]. In the QASM language, a quantum program is described as a linear sequence of gate instructions. For example, the EPR pair creation circuit is written as shown in Figure 5.1.

Due to its root from quantum circuits, sequential QASM language suffers from its limitation on modeling complex classical control, such as "repeat-until-success" procedure and other non-trivial branching. To remedy this, and to improve its expressiveness and coverage, a number

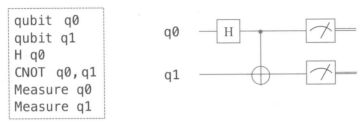

Figure 5.1: The QASM code and circuit diagram for creating an EPR pair with measurements.

of extensions to QASM have been developed. In these extensions, basic constructs (commonly used in classical programming) such as loops, subroutine calls, barriers, and classical feedback control are added. For example, the OpenQASM [163] backend has been developed by IBM Q, and ARTIQ [164] by the trapped ion community.

Many have argued for a more expressive language to support full control over the physical properties of the machine, such as pulse features. For example, OpenPulse [165] developed by IBM is one of the efforts in that direction. OpenPulse is a set of tools for building experiments out of pulses. The performance of the experiments replies heavily on the programmer's understanding of the physical system.

These low-level languages are naturally more closely tied to hardware specifics. Hence, optimizations for compiling to such languages must be tailored to the specific characteristics of their supported hardware, including device topology and noise rates etc. An efficient low-level software tool can allow quantum algorithms to be successfully executed on resource-constrained machines, such as NISQ computers.

5.2   HIGH-LEVEL PROGRAMMING LANGUAGES

A high-level programming language is needed to represent complex classical and quantum information processing in quantum algorithms. One notable example that we will return to later is the representing and compiling for hybrid classical-quantum computations.

Designing a language that enables programmers to exploit these quantum properties on real hardware while maintaining usability remains challenging. Balancing between abstraction and detail is key. On one hand, exposing device specifics helps programmers write more efficient code, but on the other hand it dramatically increases the complexity of the language. In this section, we provide a short review of the rapid development of QC languages and software tools in recent years from both academia and industry.

Recall from the beginning of this chapter, we argue that the unique properties, such as the probabilistic, entangling, no-cloning, and error-prone nature of quantum states, influence the execution model of quantum programs, and hence the design of a semantically safe programming language. Due to its complexity, a QC system requires fast and reliable classical control. We encourage the reader to revisit Chapter 1 for details about the classical-quantum co-processor model. Due to this hybrid nature of classical and quantum information processing, most existing quantum programming languages are themselves Domain-Specific Languages (DSLs). The common approach is to either directly embed in or follow similar design of a classical base language. The benefits are apparent here—instead of writing an entirely new language and its software ecosystem, we can reuse parts of a widely used language and inherit some of its optimizations such as resolving control flow. Furthermore, programmers do not need to learn a new language from scratch. Nonetheless, many hope that a new language that follows more rigorously the theory of quantum information processing will aid the discovery of new algorithms. Some have proposed new data typing for qubits and a strongly typed language to ensure proper manipulations of qubits.

Just as with classical programming languages, quantum programming languages can be classified in to two categories: functional and imperative. A functional language encourages more mathematical, abstract, compact implementation of algorithms. Examples of functional quantum programming languages include Quipper [166], Quafl [167], LIQuI|⟩ ("Liquid") [168], and Q# [169]. An imperative language describes the steps of algorithms sequentially in greater detail. It allows direct modifications of variables and often is more resource efficient. Examples of imperative quantum programming languages include Scaffold [170] embedded in C/C++, and ProjectQ [171] and Quil [172] embedded in Python.

Current NISQ systems are rapidly evolving and are highly resource constrained. Any language (together with its compilation software) will need to be versatile enough to keep up with the fast rate of change in QC systems. For example, NISQ applications such as Variational Quantum Eigen-solver (VQE) require multiple rounds of interleaved classical-quantum processing, which presents new challenges in language design and compilation optimizations.

Compared to hardware development and theoretical algorithmic understanding, our experience with the QC programming language design and its supporting software toolchain is rather nascent. With recent pushes of cloud-based access to QC hardware in the industry (such as [84, 158, 173, 174] and more to come), more and more realize the need for a full-stack QC software and hardware. We expect the growing developer community on quantum computing will pay more attention to the design of quantum programming languages and their software toolchain.

5.3   PROGRAM DEBUGGING AND VERIFICATION

How do we know a quantum program implements the transformation as intended? Can we prevent programmers from writing code that violates some quantum properties? Verification of quantum programs is a unique and non-trivial task, but we take a moment to discuss some exciting recent developments in program testing inspired by classical techniques, such as program debugging, formal logic, and proof assistant.

There are two possible notions of verification in quantum computation: *hardware verification* and *software verification*. The former refers to the problem of verifying that hardware is capable of performing quantum logic operations as intended by a program. The latter refers to the problem where we want to verify that a quantum program is bug-free and implements the desired transformation.

- *Hardware Verification.* We need tools to understand and characterize the machines that we build. At a basic level, the behavior of quantum devices can be characterized through a process termed *quantum tomography* (see reviews in [175]), where multiple measurements are used to estimate quantum states. As machines become larger, however, a systems-level approach is needed. One possible approach would be to compute and uncompute a circuit and use tomography to determine whether a machine returns to its initial state. More sophisticated tests attempt to measure the "quantumness" of a machine and its ability to create entanglement across its qubits. Validating quantumness in a machine using a (possibly purely classical) prover is a challenging task and is tightly related to computational complexity theory. For the lack of space, this book will omit some of the seminal work in verifying quantum hardware. We refer the interested readers to the Summary section at the end of the chapter, as well as reviews in [176–178].

- *Software Verification.* The purpose of software verification is typically two-fold: verifying that (i) *high-level programs* are bug-free, and that (ii) *compiler transformations* preserve logical equivalence. If we have a simulator or working machine, we can perform end-to-end unit tests or we can invest some extra quantum bits to test assertions. Methods [175] have been developed to test for basic properties such as whether two quantum states are equal, whether two states are entangled, or whether operations commute. We can either adopt testing-based or formal-methods approaches for this problem (or hybrids of both). The rest of the section is devoted to detailed discussions on verifying quantum software.

Three useful verification approaches that are being widely used today include the application of *classical simulation*, *quantum property testing*, and *formal logic*. Although these techniques do not prevent/detect all types of errors nor do they scale well to large systems, they are found to be useful in partially verifying aspects of the computing process so as to gain some confidence of its success rate. Both software and hardware verifications are expected to be particularly important for NISQ computers, shielding against the adverse impacts of errors from buggy programs, unreliable compilers, or noisy hardware.

5.3.1   TRACING VIA CLASSICAL SIMULATION

The most widely used verification approach is arguably tracing the evolution of quantum states using classical simulation. Both software and hardware verification problems can be tackled using classical simulation, because it allows us to exactly compute and compare the inputs and outputs of an ideal transformation with those of an actual one. Simulation is informative, in that it allows us to do code tracing and reveals the states of a quantum program step by step. Classical simulation, however, is not easy, as it requires to compute transformations in an exponentially large state space. Interestingly, there exists a fundamental tension between classical simulation of quantum computations and quantum supremacy. If we can efficiently simulate quantum computation on a classical computer, then we have proven that this quantum computation does not demonstrate quantum supremacy! Verification approaches involving too much of an algorithm's state space also have similar implications. If we are optimistic and assume that some quantum algorithms have supremacy over classical algorithms, then we must come up with restricted verification properties that only require partial simulation or formal verification of a sub-exponential state space. For readers who are not familiar with classical simulation techniques, please see Chapter 9 for details.

Classical simulation is a useful lens toward the subtle boundary of quantum supremacy [146, 155, 179]. Classical computers can simulate quantum algorithms consisting of only "Clifford gates" in time polynomial in the number of qubits used in the algorithm, which proves that these algorithms do not demonstrate quantum supremacy. Algorithms such as Shor's factoring algorithm, which are exponentially better than known classical algorithms, contain T gates as well as Clifford gates. We do not know how to classically simulate Shor's algorithm in sub-exponential time. We do know, however, how to simulate algorithms consisting of Clifford gates and T gates in time polynomial in the number of qubits and exponential in the number of T gates [180]. So practically, we *can* afford circuits with not too many T gates in classical simulation. That is not completely uncorrelated with the fact that there are resource theories defined around the number of T gates to help us understand the boundary between classical and quantum computing power. Verification through simulation might exploit the classical side of this boundary by trying to define correctness properties that only require simulation of parts of an algorithm that contain a small number of T gates. What these properties will be, however, is very much an open area of research.

But even classical simulation does not capture all sources of errors. In order to completely imitate the quantum process, we must include the impacts of hardware noise. The difficulty for simulation with noise is two-fold—first our understanding of the physical noise today is still limited. Modeling realistic noise remains an active field of research; second, even if we have a perfect noise model, noise simulation itself is extremely challenging. No known efficient methods exist that accurately simulates the effects of noise but scales sub-exponentially in the number of qubits.

5.3.2   ASSERTION VIA QUANTUM PROPERTY TESTING

Here we describe how to perform quantum assertions (for instance, testing whether two quantum state are equal) in quantum circuits.

Property testing, also known as hypothesis testing,[1] is an area that aims to design algorithms to check some (global) properties are present or absent in some large object through restricted (local) queries to the object. More concretely, following [175], we have the next definition:

Definition 5.1   A *property* $\mathcal{P}$ for a set of objects $\mathcal{X}$ is a subset of $\mathcal{X}$, that is, $\mathcal{P} \subseteq \mathcal{X}$. Let $d : \mathcal{X} \times \mathcal{X} \to [0, 1]$ be a distance measure on $\mathcal{X}$.

- An object $x \in \mathcal{X}$ is ϵ-*far* from $\mathcal{P}$ if $d(x, y) \le \epsilon$ for all $y \in \mathcal{P}$.

- An object $x \in \mathcal{X}$ is ϵ-*close* to $\mathcal{P}$ if there exists $y \in \mathcal{P}$ such that $d(x, y) \ge \epsilon$.

Definition 5.2   An algorithm is an ϵ-*property tester* of $\mathcal{P}$ if it accepts $x \in \mathcal{X}$ with probability of at least $2/3$ if $x \in \mathcal{P}$ or rejects $x \in \mathcal{X}$ with probability of at least $2/3$ if x is ϵ-far from $\mathcal{P}$.

Quantum property testing extends the definitions to use quantum algorithms to test quantum objects.

Testing Properties of Quantum States

Here we briefly describe some of the widely used strategies for discriminating quantum states. For a d-dimensional pure state $\psi \in \mathbb{C}^d$, a convenient distance measure is the trace distance:

$$D_{\mathrm{tr}}(|\psi\rangle, |\phi\rangle) = \frac{1}{2} |||\psi\rangle\langle\psi| - |\phi\rangle\langle\phi|\,||_1 = \sqrt{1 - |\langle\psi|\phi\rangle|^2},$$

where $||\cdot||_1$ is the 1-norm.

Ideally, we want to find an algorithm that tests for a property (that is a ϵ-tester) using a small number of copies only in terms of ϵ, regardless of d. When this is not possible, we attempt to minimize the dependency on d.

In the following we give some example properties of quantum states and illustrate their property testers.

- *Testing if a state $|\psi\rangle$ is equal to another known state $|\phi\rangle$.* Note that "equal" here means that the two states are the same up to global phase, that is satisfying $|\psi\rangle = e^{i\theta} |\phi\rangle$ for some real number θ. The simplest test for this property is to measure $|\psi\rangle$ in the basis $\{|\phi\rangle\langle\phi|, I - |\phi\rangle\langle\phi|\}$ and accept if the first outcome is observed. Notice that the acceptance probability is exactly $|\langle\psi|\phi\rangle|^2$ which is 1 if the two states are equal; the rejection probability is $1 - |\langle\psi|\phi\rangle|^2$ which is ϵ^2 if the trace distance $D_{\mathrm{tr}}(|\psi\rangle, |\phi\rangle) \ge \epsilon$. So we can repeat the test and verify equality with $O(1/\epsilon^2)$ copies. *We remark that, in fact, any non-trivial properties on pure states require $\Omega(1/\epsilon^2)$ copies to achieve the desired $2/3$ success probability.*

[1]This is not to be confused with statistical hypothesis testing which is a method of statistical inference based on sampling data sets. However, the two concepts are not without any connections.

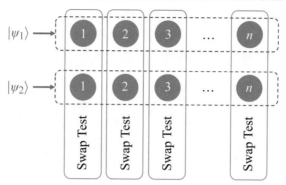

Figure 5.2: Testing productness of $|\psi\rangle$ using a series of swap tests on $|\psi\rangle \otimes |\psi\rangle$.

- *Testing if two unknown (possibly mixed) states, ρ and σ, are equal.* To test this property, we introduce an important procedure called the *swap test*, due to Buhrman et al. [181]. Take two quantum states ρ and σ and an ancilla qubit $|0\rangle$.

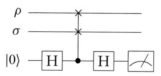

In this circuit, a controlled-swap gate is sandwiched by two Hadamard gates. When we measure the third register (ancilla qubit), we accept if the outcome is $|0\rangle$, and reject if the outcome is $|1\rangle$. In essence, this is a "similarity test," because one can derive [181, 182]:

- (Pure state case) $\mathbf{Pr}[\text{accept} |\psi\rangle \otimes |\phi\rangle] = \frac{1}{2} + \frac{1}{2} |\langle\psi|\phi\rangle|^2 = 1 - \frac{1}{2} D_{\text{tr}}(|\psi,|\phi\rangle))^2$.
- (Mixed state case) $\mathbf{Pr}[\text{accept}\rho \otimes \sigma] = \frac{1}{2} + \frac{1}{2}\text{tr}(\rho\sigma)$.

We can analyze similarly for the probability of rejection (see details in [175]) and obtain the tester using $O(1/\epsilon^2)$ copies. The swap test is also optimal for testing such a property. One can generalize to equality test for multiple states with techniques such as permutation tests [183].

- *Testing if a pure state $|\psi\rangle$ is an entangled state.* In particular, we want to test if $|\psi\rangle$ can be written as a tensor product of n local states (i.e., speedup $|\psi\rangle = |\psi_1\rangle \otimes |\psi_2\rangle \otimes \cdots \otimes |\psi_n\rangle$ or not (in which case, $|\psi\rangle$ is entangled). We can use a series of swap tests on two copies of $|\psi\rangle$ and repeat for $O(1/\epsilon^2)$ times to test for productness [184, 185]. More specifically, the swap test is applied to each pair of the n local parts in $|\psi\rangle \otimes |\psi\rangle$, as shown in Figure 5.2.

Many other testers exist for properties of quantum states. For example, we can test if a state belongs to stabilizer states [157, 186], matrix product states [187], pure vs. mixed states [188], low Schmidt-rank states [189], or having arbitrary finite properties [190].

Testing Properties of Quantum Dynamics

One can also test properties of some quantum transformations, where we are given a black-box transformation U and need to decide whether U has some property or is far from having it. Tests vary depending on assumptions such as whether we are given access to the controlled operator $(c - U)$ and/or the inverse operator (U^{-1}) in addition to U itself.

For unitary operators, two common distance measures are used to evaluate the performance of the tests: one for worst-case analysis and the other for average-case analysis. Here we take the definitions over pure input states as examples (see Chapter 8 for generalizations to mixed states).

- For two d-dimensional unitary operators U, V, we define the worst-case distance over all possible pure states as:

$$D_{\max}(U, V) = \max_{|\psi\rangle} D_{\mathrm{tr}}(U\,|\psi\rangle - V\,|\psi\rangle) = \max_{|\psi\rangle} \sqrt{1 - |\langle\psi|U^\dagger V|\psi\rangle|^2}.$$

- For two d-dimensional unitary operators U, V, we define the average-case distance as:

$$D_{\mathrm{avg}}(U, V) = \frac{1}{\sqrt{2}}||A \otimes A^\dagger - B \otimes B^\dagger||_2 = \sqrt{1 - |\langle U, V\rangle|^2},$$

where $||M||_2 = \sqrt{\frac{1}{d}\sum_{i,j=1}^d |M_{ij}|^2}$ is the 2-norm, and $\langle U, V\rangle = \frac{1}{d}\mathrm{tr}(U^\dagger V)$ is the Hilbert–Schmidt inner product.

One useful tool that maps properties of quantum states to properties of unitaries is called the *Choi–Jamiołkowski isomorphism* [191, 192]. This tool sometimes allows us to take a test for properties of quantum states and apply it directly to test for properties of unitary operators. The idea is to first prepare the maximally entangled state of two d-dimensional systems

$$|\psi\rangle = \frac{1}{\sqrt{d}}\sum_{i=1}^d |i\rangle\,\langle i|$$

and then apply the unitary operator U to the first system:

$$|U\rangle = \frac{1}{\sqrt{d}}\sum_{i,j=1}^d U_{ji}\,|j\rangle\,\langle i|.$$

Now applying tests on the two states $|U\rangle$ and $|V\rangle$, we have equivalently obtained tests for U and V. Some example properties where we can use the Choi–Jamiołkowski isomorphism include: equality of U and V, and U being a product operator, etc.

One may also find other tests for (Pauli/Clifford) group membership [190, 193] and commutativity [194] interesting.

5.3.3   PROOFS VIA FORMAL VERIFICATION

Some progress has also been made in applying formal methods to verify quantum computations. Quantum programs are typically implemented with well-defined semantics. We can usually deduct the behavior of quantum circuits directly from their descriptions. QWire [195] uses Coq [196] to verify some properties of simple quantum circuits, but classical computation for the theorem prover scales exponentially with the number of qubits. Feynman-path sum technique has recently been used to efficiently test for circuit equivalence [197]. Once again, the key challenge is to define useful correctness properties that a theorem prover can handle more scalably. Quantum Hoare logic [198, 199], implemented in Isabelle, is a program logic that simplifies the full verification of programs. Some tools are designed to verify specific sub-classes of quantum circuits. For instance, relational quantum Hoare logic [200] is developed for security protocols. ReVerC [201] targets reversible circuits and is verified with the proof assistant F*.

5.4   SUMMARY AND OUTLOOK

Some may hope that the "right" quantum programming language will facilitate the development of many more novel quantum algorithms, but quantum algorithms have thus far been developed with pen and paper using mathematical techniques. More realistically, quantum programming languages are essential in converting theoretical descriptions of algorithms to practical implementations that are both correct, efficient, and adapted for specific applications. This process can take months of effort for each algorithm and application.

Quantum programming languages are the first set of abstractions that we have approached in this book, and they help set the framework for optimization and verification. In subsequent chapters we shall see that these abstractions are still fluid as quantum computer systems develop and we explore which cross-layer optimizations are critical for efficiency.

Further Reading

The chapter has mostly discussed quantum programming languages developed around the circuit model of quantum computation. These languages represents quantum circuits at multiple levels of abstraction, and offer powerful tools for circuit optimization. Among them, some have device-driven designs, including OpenQASM [163] and Quil [172]; others have algorithm-driven designs, such as Q# [169].

In the case of functional programming languages, the lambda calculus model for quantum computation [202, 203] serves as their theoretical foundation. In particular, lambda calculus has motivated a series of development in type systems for quantum computation, which result in proofs on quantum data and quantum functions [166, 195].

Some recent developments in quantum programming languages have escaped the notion of quantum circuit; rather, they model a quantum process by undirected graphs, such as the ZX-diagrams. These graphical models lead to languanges such as ZX-calculus [204, 205] and ZW-calculus [206]. The ZX-diagrams are closely related to the tensor network representation and undirected graphical models introduced in Chapter 9; we encourage the interested reader to make the connection.

CHAPTER 6

Circuit Synthesis and Compilation

Practical quantum computation may be achievable in the next few years, but applications will need to be error-tolerant and make the best use of a relatively small number of quantum bits and operations. Compilation tools will play a critical role in achieving these goals. The job of a quantum compiler is to translate a quantum program written in a high-level programming language into native instructions recognizable by the hardware, through a series of transformations and optimizations. Traditional wisdom from compilation for classical computers can occasionally be inherited or adapted to the quantum case, such as resolving control flows and allocating registers. This chapter puts particular emphasis on the aspects of compilation that are unique to quantum computers. Notably, compilation under strict resource constraints has proven challenging, and optimization will have to break traditional abstractions and be customized to algorithmic and device characteristics in a manner never before seen in classical computing. We call attention to a number of important steps specialized for quantum compilation to help ensure the efficiency and correctness needed. To name a few, unitary synthesis focuses on exactly or approximately expressing arbitrary unitary transformations (such as single qubit rotations by an arbitrary angle) in a sequence of elementary gates. The goal of gate scheduling is to utilize commutation relations to determine the ordering of the (possibly parallel) operations, and to use circuit equivalence to simplify quantum programs. Qubit mapping is another challenge, in that we aim to strategically assign the variables in a quantum program to the qubits available in the system, under multiple constraints such as limited connectivity between qubits, fluctuations in the reliability of qubits and links, and potential opportunities for reclamation and reuse of qubits, etc.

6.1 SYNTHESIZING QUANTUM CIRCUITS

This section aims to address one essential question in quantum compiling, namely how to (efficiently) implement some arbitrary unitary transformation using a given finite set of realizable quantum gates (i.e., primitive instructions). The complexity of the problem varies, depending on the objectives and assumptions. As a result, efficiency and optimality of the solutions varies. So it is important to recognize the different situations being considered in the community, and categorize the known synthesis techniques into classes accordingly. Some example types of synthesis techniques considered in the rest of the section can be summarized as follows:

- choice of universal instruction set;

- single-qubit, multi-qubit, and qudit (i.e., d-level quantum logic) synthesis; and

- exact and approximate synthesis.

The existence of an efficient synthesis of quantum circuits is largely determined by the choice of instruction set; one can imagine some quantum gates to be more "powerful" than others. Here powerful, which will be defined in subsequent sections, can be informally thought of as being able to cover the entire space of possible unitary gates more quickly. Furthermore, synthesis for multi-qubit unitaries or qudit unitaries is believed to be more difficult in general due to the high dimensionality involved; a general strategy is to decompose the high-dimensional unitary matrices into pieces of one- or two-qubit unitary matrices, for which efficient synthesis methods are known. Last, strategies for exactly or approximately synthesizing quantum circuits differ significantly, and consequently, they can have drastically different complexity.

For example, if a quantum program is to be executed on a superconducting NISQ computer (without quantum error correction), then the instruction set would likely consist of single-qubit rotation ($R(\theta)$) gates and two-qubit cross-resonance (CR) gate, for they are easier to implement to high precision. Consequently, the target transformation U of the quantum program is synthesized, exactly or approximately (depending on the precision tolerance). The synthesis procedure typically involves first decomposing the multi-qubit unitary U into sequence of single-qubit unitaries and two-qubit CR gates, and then decomposing the single-qubit unitaries into Pauli rotation gates. The goal would be to synthesize the most efficient circuit (e.g., short in depth and small in number of qubits) to some high precision required by the target computer.

We have just demonstrated a typical example in circuit synthesis; the rest of the section focuses on illustrating systematically how to realize arbitrary quantum circuits under various conditions.

6.1.1   CHOICE OF UNIVERSAL INSTRUCTION SET

The first step in synthesizing quantum circuits is to choose a *computationally universal* quantum instruction set. What does it mean for a quantum instruction set S to be universal?

Definition 6.1   A quantum instruction set S is called *computationally universal* if and only if gates from S allow the realization of an arbitrary quantum circuit C.

One might worry that such realization could consist of a large number of gates from S. Fortunately, efficient universality results can obtained for some instruction sets, which is the focus of the section.

One way of demonstrating that a given instruction set is universal is by providing a constructive algorithm for *exactly* decomposing an arbitrary unitary transformation into a product of gates from the instruction set. It is convenient, and sometimes equally valuable, to relax the constraint on exact synthesis, and show a product of gates from the instruction set that is *very close to* the target unitary transformation, i.e., a universal realization up to some precision. In the following, we show that one- and two-qubit unitary gates are universal by exactly synthesizing an arbitrary n-qubit unitary transformation.

An arbitrary single-qubit unitary transformation can be mathematically represented as a 2×2 matrix:

$$\begin{pmatrix} e^{i\alpha}\cos(\theta) & e^{i\beta}\sin(\theta) \\ -e^{-i\beta}\sin(\theta) & -e^{-i\alpha}\cos(\theta) \end{pmatrix}$$

which is parameterized by continuous variables α, β, θ. More generally, the set of all possible n-qubit gates is the determinant-1 unitary transformations on a 2^n-dimensional vector space, denoted by the group $SU(2^n)$. In this notation, one can rewrite the definition of computational universality as follows.

Theorem 6.2   *One-qubit and two-qubit unitary gates are universal; an arbitrary n-qubit unitary transformation U in SU(d), where d $= 2^n$, can be represented as a product of O(2^{2n}) matrices of the block form:*

$$\Gamma_i(V) = \begin{pmatrix} I_i & 0 & 0 \\ 0 & V & 0 \\ 0 & 0 & I_{d-i} \end{pmatrix},$$

where V $\in$ SU(2) is a 2 $\times$ 2 matrix, and I_k is a k $\times$ k Identity matrix.

Each $\Gamma_j(V)$ can be thought of as a two-qubit gate on the ith and the $(i+1)$th qubits, while keeping the rest untouched. We explain how a target unitary transformation acting on n qubits is decomposed into a product of two-qubit unitaries. In this construction, we find a sequence of suitable matrices $W_1, W_2, \ldots, W_{2^n}$ such that

$$W_{2^n} \cdots W_2 W_1 = U.$$

It is equivalent to write

$$W_{2^n} \cdots W_2 W_1 U^{-1} = I.$$

The key in this construction is that each W_j, for $j \in [2^n]$, transform the jth column of U^{-1} to the jth column of I. W_j is accomplished by a product of $\Gamma_i(V)$, for $i \in [2^n]$, where each $\Gamma_i(V)$ transforms the ith and $(i+1)$th qubits. Since $\Gamma_i(V)$ only operates on the subspace of two qubits, it is sufficient to see that there exists matrix V such that

$$V\begin{pmatrix} a \\ b \end{pmatrix} = \begin{pmatrix} \sqrt{|a|^2 + |b|^2} \\ 0 \end{pmatrix}$$

for any number $a, b \in \mathbb{C}$. Indeed, we can write down such matrix:

$$V = \begin{pmatrix} \dfrac{a^*}{\sqrt{|a|^2+|b|^2}} & \dfrac{b^*}{\sqrt{|a|^2+|b|^2}} \\ \dfrac{b}{\sqrt{|a|^2+|b|^2}} & \dfrac{-a}{\sqrt{|a|^2+|b|^2}} \end{pmatrix}.$$

This is therefore an algorithm for *exactly* decomposing an arbitrary unitary U into a product of $O(2^n \cdot 2^n)$ matrix of the form of $\Gamma_i(V)$, each of which acts only on a two-qubit subspace. It is also important to note that, although this is an exact synthesis method, the precision of complex numbers δ will impact the final synthesis precision, which adds a $\text{poly}(\log(1/\delta))$ factor to the final algorithm complexity.

It is also important to note that two-qubit unitary gates are *necessary* for universal quantum computation, as single-qubit gates alone cannot produce the entanglement between qubits required by almost all interest quantum algorithms.

Theorem 6.2 has significant practical implications. In principle, a universal quantum computer can be realized by implementing one- and two-qubit unitary gates to high precision. In practice, directly implementing a unitary transformation $U \in SU(2^n)$ would require exponentially complex control mechanism for all the parameters in the large $O(2^{2n})$ matrix, while unitary synthesis techniques allow one to realize an equivalent transformation using only control mechanism that acts on only one or two qubits at a time.

In particular, for experimentalists, much engineering efforts have been put into physically implementing single-qubit gates and CNOT gates on qubits. This is because, as noted in Chapter 2, single-qubit gates and CNOT gate are universal. The proof is omitted here, but can be found in [86]. The key idea is to show that single-qubit gates and CNOT gate can be used to make up all two-qubit unitaries. Thus, following Theorem 6.2, we can show that any quantum circuit made of n-qubit gates can be exactly simulated by single-qubit gates and CNOT gates, with only a linear increase in the number of gates.

From an algorithmic/number-theoretic perspective, choosing a universal instruction set with the most convenient structure is preferred. As we will see in the following section, highly structured instruction sets yield fast and efficient synthesis algorithms. Some examples of structured instruction sets [207–211] include the Clifford-T set, the Clifford-cyclotomic set, and the V-basis set, etc.

6.1.2   EXACT SYNTHESIS

Following the previous section where we showed a constructive algorithm for exactly synthesizing an arbitrary unitary for $SU(d)$, we continue to provide some exact synthesis examples. The hope is that with prior knowledge in the structures of a target unitary transformation, we can find a more efficient circuit implementation. Indeed, much progress has been made in specialized synthesis algorithm, we focus our discussions on some commonly used transformations, such as controlled unitary gates $\Lambda_k(U)$ (where U is performed on the target qubit(s) conditioned on the state of the k controlled qubit(s)), quantum oracle gates O_f (where a classical reversible function f is computed on the quantum state), and rotation gates $R_\alpha(\theta)$ (where a θ angle is rotated about an axis α in the Hilbert space of the quantum state). Last, we briefly demonstrate exact synthesis to Clifford+T gates, as an example technique where the structure of the instruction set is relatively well understood.

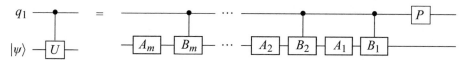

Figure 6.1: A generic decomposition for controlled unitary gate ($\Lambda(U)$). Note that P is the phase shift gate $P = |0\rangle \langle 0| + e^{i\alpha} |1\rangle \langle 1|$.

It is important to note that Clifford+T synthesis originates from compilations for fault-tolerant (FT) machines, where Clifford operations are easy to realize using the stabilizer formalism (see details in Chapter 8). An exciting open direction is to define, for NISQ computers, an instruction set with rich structures and efficient synthesis algorithms.

Synthesizing Controlled-Unitary

Recall that a controlled unitary gate (namely $\Lambda(U)$ or $c\text{-}U$) acts on $1 + t$ qubits, 1 of which is the controlled qubit and t of which are the target qubits:

$$\Lambda(U) |q_1\rangle \otimes |\psi\rangle = |q_1\rangle \otimes U^{q_1} |\psi\rangle = \begin{cases} |q_1\rangle \otimes U |\psi\rangle & \text{if } |q_1\rangle = |1\rangle, \\ |q_1\rangle \otimes |\psi\rangle & \text{if } |q_1\rangle = |0\rangle. \end{cases}$$

More generally, a multi-controlled-unitary gate acts on $(k + t)$ qubits, k of which are the controlled qubits and t of which are the target qubits:

$$\Lambda_k(U) |q_1 \cdots q_k\rangle \otimes |\psi\rangle = |q_1 \cdots q_k\rangle \otimes U^{q_1 \cdots q_k} |\psi\rangle$$
$$= \begin{cases} |q_1 \cdots q_k\rangle \otimes U |\psi\rangle & \text{if } |q_1 \cdots q_k\rangle = |1 \cdots 1\rangle, \\ |q_1 \cdots q_k\rangle \otimes |\psi\rangle & \text{otherwise,} \end{cases}$$

where $q_1 \cdots q_k$ in the exponent of U is the product of the bits $q_1, \ldots, q_k$.

The general strategy to synthesizing controlled unitary gates ($\Lambda(U)$) is to rewrite unitary U in a product form:

$$U = e^{i\alpha} A_m B_m \ldots A_2 B_2 A_1 B_1$$

satisfying $A_m \cdots A_2 A_1 = I$ and B_i are chosen such that $\Lambda(B_i)$ are easier to implement than $\Lambda(U)$ itself, e.g., $B_i = X$ so that $\Lambda(X) =$ CNOT gate. Here α is a phase factor, and m is typically to be minimized. The key idea is that now the controlled unitary $\Lambda(U)$ is decomposed as $\Lambda(U) = \Lambda(e^{i\alpha})(I \otimes A_m)\Lambda(B_m)\ldots(I \otimes A_2)\Lambda(B_2)(I \otimes A_1)\Lambda(B_1)$, that is a sequence of interleaved non-controlled unitary gates and controlled unitary gates. Note that $\Lambda(e^{i\alpha})$ denotes the controlled phase shift gate: $\Lambda\left(\begin{pmatrix} e^{i\alpha} & 0 \\ 0 & e^{i\alpha} \end{pmatrix}\right) = \begin{pmatrix} 1 & 0 \\ 0 & e^{i\alpha} \end{pmatrix} \otimes I$. It can be visualized as the circuit in Figure 6.1.

It has been shown that an arbitrary unitary U can be decomposed into $U = e^{i\alpha} CXBXA$, where X is the Not gate and $CBA = I$ [86]. As a result, $\Lambda(U)$ is implemented using single-qubit

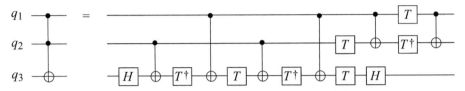

Figure 6.2: An quantum circuit implementation of the reversible Toffoli gate as a sequence of single- and two-qubit gates.

gates and CNOT gates. Since the choice of A, B, C is non-unique, one often needs to further optimize the synthesis of $\Lambda(U)$ by intelligently choosing a decomposition of U.

Similar logic applies to multi-controlled unitary gate $(\Lambda_k(U))$ [86]. An important example is the Toffoli gate. Figure 6.2 shows a synthesized Toffoli gate using Hadamard gates, T gates (and their inverses), and CNOT gates.

Synthesizing Quantum Oracles

Quantum oracles are essential components in quantum query algorithms; they allow one to compute classical Boolean functions on quantum states. Implementations of quantum oracles have been largely under-studied, partially because a traditional analysis of the query complexity of an quantum algorithm typically assumes that oracles are given by another party, but aims to bound the number of accesses made to the oracles. However, in practice, oracles need to be implemented just like the rest of the algorithm. The cost of the oracles eventually contributes to the time cost of the overall algorithms. As such, this section is devoted to general strategies for synthesizing quantum oracles.

Do all classical Boolean functions have quantum circuit implementations? For those that do, can we efficiently synthesize them? The basic principles introduced in Chapter 2 give rise to the potential computing power that quantum computers possess, but at the same time, they impose strict constraints on what we can do in quantum computation. For example, the transformation rule implies that any quantum logic gate we apply to a qubit has to be *reversible*. The classical AND gate in Figure 6.3 is *not* reversible because we cannot recover the two input bits based solely on one output bit. To make it reversible, we could introduce a scratch bit, called *ancilla*, to store the result out-of-place, as in a controlled-controlled-NOT gate (or Toffoli gate) in Figure 6.3. As the arithmetic complexity scales up when tackling difficult computational problems, we quickly see extensive usage of ancilla bits in our circuits due to this *reversibility constraint*.

We first demonstrate an example where the oracle is simple to synthesize, namely, the phase oracle for the Berstein–Vazirani algorithm from Chapter 3; ultimately, we will show that reversible logic synthesis tools are necessary for more complex functions.

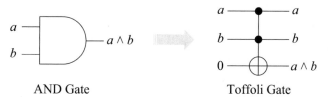

AND Gate   Toffoli Gate

Figure 6.3: Circuit diagram for the irreversible AND gate and the reversible Toffoli gate.

Recall that, in the Berstein–Vazirani algorithm, a phase oracle implements a Boolean function $f : \{0,1\}^n \to \{0,1\}$, which encodes a secret string $s \in \{0,1\}^n$ as follows:

$$f(x) = s \cdot x = \sum_{i=1}^{n} s_i \cdot x_i \quad \mathrm{mod}\ 2.$$

When applied to a quantum state $|\psi\rangle$, the oracle $O_f^{\pm}$ accumulates an phase on $|\psi\rangle$ depending on the output of f. Without loss of generality, we consider a basis of $|\psi\rangle$, denoted as $|q_1 q_2 \ldots q_n\rangle$.

$$O_f^{\pm} |q_1 q_2 \ldots q_n\rangle = (-1)^{f(q_1 q_2 \ldots q_n)} |q_1 q_2 \ldots q_n\rangle.$$

The key in synthesizing an oracle $O_f^{\pm}$ for a particular secret string $s \in \{0,1\}^n$ is by a technique called *phase kickback*.

Phase Kickback

In Chapter 2 we demonstrated that a CNOT gate flips the state of the target bit, conditioned on the control bit. In the following, we will see that the control bit is sometimes affected by the CNOT gate as well. For example, we have the following *phase kickback* circuit:

$$\alpha|0\rangle + \beta|1\rangle \quad\quad \alpha|0\rangle - \beta|1\rangle$$
$$|-\rangle \quad\quad |-\rangle$$

Normally the state of the control bit does not change after a CNOT gate. Here if the target bit is in the $|-\rangle$ state, as the name of the circuit suggested, a phase (in this case, a minus sign) is "kicked" onto the control bit.

The control qubit is in arbitrary state $\alpha|0\rangle + \beta|1\rangle = \begin{pmatrix} \alpha \\ \beta \end{pmatrix}$, and the target qubit is

$|-\rangle = \frac{1}{\sqrt{2}}(|0\rangle - |1\rangle) = \begin{pmatrix} \frac{1}{\sqrt{2}} \\ -\frac{1}{\sqrt{2}} \end{pmatrix}$. So the 2-qubit system has a joint state:

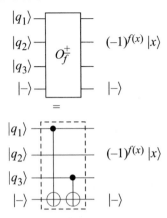

Figure 6.4: Oracle implementation for Bernstein–Vazirani algorithm with secret string $s = 101$.

$$\begin{pmatrix} \alpha \\ \beta \end{pmatrix} \otimes \begin{pmatrix} \frac{1}{\sqrt{2}} \\ -\frac{1}{\sqrt{2}} \end{pmatrix} = \frac{1}{\sqrt{2}} \begin{pmatrix} \alpha \\ -\alpha \\ \beta \\ -\beta \end{pmatrix}.$$

Applying the CNOT gate, we arrive at the state:

$$\begin{pmatrix} 1 & 0 & 0 & 0 \\ 0 & 1 & 0 & 0 \\ 0 & 0 & 0 & 1 \\ 0 & 0 & 1 & 0 \end{pmatrix} \cdot \frac{1}{\sqrt{2}} \begin{pmatrix} \alpha \\ -\alpha \\ \beta \\ -\beta \end{pmatrix} = \frac{1}{\sqrt{2}} \begin{pmatrix} \alpha \\ -\alpha \\ -\beta \\ \beta \end{pmatrix} = \begin{pmatrix} \alpha \\ -\beta \end{pmatrix} \otimes \begin{pmatrix} \frac{1}{\sqrt{2}} \\ -\frac{1}{\sqrt{2}} \end{pmatrix}.$$

Remarkably, by performing the CNOT gate, we have essentially changed the state of the control qubit, i.e., adding a phase to it. Therefore, the CNOT gate accomplishes a phase kickback.

Now with phase kickback, we remark that synthesizing $O_f^{\pm}$ becomes straight-forward: for every bit b_i in string s, if $b_i = 1$, we add a CNOT gate between the $|q_i\rangle$ and the bottom qubit $|-\rangle$, as shown in Figure 6.4.

More generally, a typical synthesizer follows a two-step procedure to generate the suitable quantum program that implements the desired reversible arithmetic.

1. Find an efficient implementation of the desired function using reversible logic gates.

2. Implement each reversible logic gate using quantum gate(s). For example, a `Toffoli` gate can be implemented with a sequence of single-qubit gates and a few `CNOT` gates, as shown in Figure 6.2.

For small arithmetic logic, algorithms exist to directly synthesize reversible circuits from the truth tables of the desired function. This typically works well for small low-level combinational functions, but not for functions with internal states [212]. As the complexity of the arithmetic in an algorithm scales up, *modularity* quickly becomes convenient, and in many cases necessary. That is, to construct high-level arithmetic, we need to start with all modular subroutines.

In reversible logic synthesis and optimization, besides making our circuit for the reversible function contain as few gates as possible, we would also like to minimize the amount of scratch memory (i.e., number of ancilla bits) used in the circuit. Fortunately, there is a way to recycle ancilla bits for later reuse. For a circuit that makes extensive use of scratch memory, managing the allocation and reclamation of the ancilla bits becomes critical to producing an efficient implementation of the function. The technique is called "uncomputation," formalized by Charles Bennett [213]. Details on uncomputation are deferred to Section 6.4.3.

To emphasize the importance of reversible logic synthesis, we remark that reversible arithmetic plays a pivotal role in many known quantum algorithms. The advantage of quantum algorithms is thought to stem from their ability to pass a superposition of inputs into a classical function at once, whereas a classical algorithm can only evaluate the function on single input at a time. Many quantum algorithms involve computing classical functions, which must be embedded in the form of reversible arithmetic subroutines in quantum circuits. For example, Shor's factoring algorithm [12] uses classical modular-exponentiation arithmetic, Grover's searching algorithm [13] also implements its underlying search problem as an oracle subroutine, and the HHL algorithm for solving a linear system of equations contains an expensive reciprocal step [214]. These reversible arithmetic subroutines are typically the most resource-demanding computational components of the entire quantum circuit.

Synthesis Over Structured Instruction Sets

When a target instruction set has well-understood, nice structures, exact synthesis of quantum circuits over such instruction set can be efficiently (and sometimes optimally) implemented. Indeed, in this section, we explain how to exploit such structures from a algebraic and number-theoretic perspective. In particular, we examine single-qubit synthesis algorithms over the *Clifford-T set*.

Recall that the single-qubit Pauli gates are denoted as

$$X = \begin{pmatrix} 0 & 1 \\ 1 & 0 \end{pmatrix}, \quad Y = \begin{pmatrix} 0 & -i \\ i & 0 \end{pmatrix}, \quad Z = \begin{pmatrix} 1 & 0 \\ 0 & -1 \end{pmatrix}.$$

Definition 6.3   The *Pauli group* P contains all quantum gates generated by $\langle X, Y, Z, i \rangle$, where X, Y, Z are the Pauli gates and i is the imaginary unit.

So any gates that can be written as a product of the Pauli group generators belong to the Pauli group, e.g., $-I = i^2 X^2$.

Definition 6.4 The single-qubit *Clifford group* C contains all single-qubit quantum gates generated by $\langle H, S, \omega \rangle$, where $H = \frac{1}{\sqrt{2}} \begin{pmatrix} 1 & 1 \\ 1 & -1 \end{pmatrix}$ is the Hadamard gate, $S = \begin{pmatrix} 1 & 0 \\ 0 & i \end{pmatrix}$ is the Phase gate, and $\omega = e^{i\pi/4}$.

Although our focus in this section is on single-qubit synthesis, we note that, for completeness, the two-qubit Clifford group can be generated by $\langle H \otimes I, I \otimes H, S \otimes I, I \otimes S, CNOT, \omega \rangle$. For example, controlled-phase (CZ) gate is one of the Clifford gates, as $CZ = (I \otimes H)CNOT(I \otimes H)$.

The Clifford gates are of particular interests for fault-tolerant quantum computers, as they are well-suited for many quantum error correction codes built upon the stabilizer formalism (please see Chapter 8 for details).

By adding the T gate, $T = \begin{pmatrix} 1 & 0 \\ 0 & e^{i\pi/4} \end{pmatrix}$, we arrive at a universal instruction set. That is to say, for single-qubit circuits, $\langle H, T, \omega \rangle$ is universal. Throughout this book, we might refer to such instruction set as the Clifford-T set.

Exact Clifford-T synthesis was first proposed in [215] and studied in [207, 208, 210, 216]. Many of the synthesis tools rely on the observation that an arbitrary (single-qubit) quantum circuit can be written in some normal form. Here we follow the analysis called the *Matsumoto–Amano normal form*.

Recall from Section 2.2, we define the Bloch sphere representation of quantum states:

$$\rho = |\psi\rangle \langle\psi| = \tfrac{1}{2}(I + xX + yY + zZ) = \begin{pmatrix} x \\ y \\ z \end{pmatrix}.$$

Definition 6.5 The *Bloch sphere representation*, denoted as $\mathcal{U} \in SO(3)$, of unitary gate U is defined as a linear operator

$$\mathcal{U} \begin{pmatrix} x \\ y \\ z \end{pmatrix} = \begin{pmatrix} x' \\ y' \\ z' \end{pmatrix},$$

satisfying $U(xX + yY + zZ)U^\dagger = x'X + y'Y + z'Z$.

For example, the Hadamard gate H can be written in this form as:

$$\mathcal{H} = \begin{pmatrix} 0 & 0 & 1 \\ 0 & -1 & 0 \\ 1 & 0 & 0 \end{pmatrix}.$$

By writing down the Bloch sphere representations for the Clifford-T generators $\langle H, S, T \rangle$, we observe that the entries in an arbitrary $\mathcal{U}$ are in the ring

$$\mathbb{Z}[\frac{1}{\sqrt{2}}] = \left\{ \frac{\sqrt{2}a + b}{\sqrt{2}^k} : a, b \in \mathbb{Z}, k \in \mathbb{N} \right\},$$

where $\mathbb{Z}$ and $\mathbb{N}$ denote the sets of integers and natural numbers, respectively. The proof of the statement can be found in [216]. In fact, if U is a single-qubit Clifford-T unitary, then its unitary matrix entries are in the ring $\mathbb{Z}[\frac{1}{\sqrt{2}}, i]$.

The Matsumoto–Amano normal form construction goes as follows—we start with any single-qubit Clifford-T unitary U, written in a generic form:

$$U = C_n T \cdots C_2 T C_1 T C_0,$$

where C_i is in the Clifford group, for $i \in \{0, 1, \ldots, n\}$. The objective is to simplify the above expression. Upon inspection of the Bloch sphere representation of the generators in the Clifford-T set, one can show that U can be uniquely rewritten in the form of

$$U = M_k \cdots M_2 M_1 C_0,$$

where $k \geq 0$, and $M_i \in \{T, HT, SHT\}$ for $i \in [k]$. Now we want to determine the M_i's iteratively.

For convenience, we write out the Bloch sphere representation of the matrices of interests:

$$\mathcal{U} = \begin{pmatrix} u_{11} & u_{12} & u_{13} \\ u_{21} & u_{22} & u_{23} \\ u_{31} & u_{32} & u_{33} \end{pmatrix}, \quad \mathcal{S} = \begin{pmatrix} 0 & -1 & 0 \\ 1 & 0 & 0 \\ 0 & 0 & 1 \end{pmatrix}, \quad \mathcal{T} = \frac{1}{\sqrt{2}} \begin{pmatrix} 1 & -1 & 0 \\ 1 & 1 & 0 \\ 0 & 0 & \sqrt{2} \end{pmatrix},$$

where the entries of $\mathcal{U}$ is defined as $u_{ij} = \frac{\sqrt{2}a_{ij} + b_{ij}}{\sqrt{2}^k}$.

The key observation is that the denominator of the entry in $\mathcal{U}$, namely the $\sqrt{2}^k$, can be obtained from a product of $\mathcal{T}$ operators, interleaved with Clifford operators. Every time we multiply the inverse of $\mathcal{T}$ on the left of $\mathcal{U}$, we reduce k by one. Specifically, we want $M_1^{-1} M_2^{-1} \cdots M_k^{-1} \mathcal{U}$ to arrive at a Clifford operator $C_0 \in \mathcal{C}$. There exists a constant-time algorithm for finding a suitable M_i for each iteration. [207] shows that $M_i \in \{T, HT, SHT\}$ for all $i \in [k]$. One can also prove that such normal form is unique [216]. Hence, this algorithm finds a Clifford-T circuit with optimal number of T gates.

Besides the Matsumoto–Amano normal form construction, there exists other synthesis tools that exploit the structure of Clifford-T circuits [208, 211, 217]. Many techniques have been developed for other similar instruction sets, such as the Clifford-cyclotomic set [209], the V-basis set [210, 218], and the Clifford-CS set [219].

So far, we have been discussing specialized synthesis tools over instruction sets that have nice algebraic structures; little is known on exact synthesis over some physically motivated instruction sets for the NISQ era. For instance, unlike a FT machine with finite set of instructions (e.g., Clifford+T gates), a NISQ machine typically supports arbitrary-angle rotations implemented via analog physical pulses with high precision, hence the single-qubit rotation gates *need not be* synthesized. It remains an open problem in finding efficient synthesis algorithms that are tied to realistic physical platforms, e.g., a noise-aware synthesizer.

6.1.3   APPROXIMATE SYNTHESIS

Suppose the unitary operator U represent the target transformation, and V is the unitary operator that is actually implemented using gates from the given set $\mathcal{G}$. V is an *approximate* implementation of the transformation U using the gate set $\mathcal{G}$.

Approximation Metrics

The *distance* of the approximate implementation V to the ideal one U is defined as:

$$D(U, V) = \sup_{|\psi\rangle} ||(U - V)|\psi\rangle||,$$

where the supremum is over all possible pure quantum states $|\psi\rangle$, and $||\mathbf{x}|| = \sqrt{\mathbf{x}^*\mathbf{x}}$ is the 2-norm of vector $\mathbf{x}$. The intuition behind this definition is that if $D(U, V)$ is small, then it is roughly equivalent to saying that the measurement outcomes of $U|\psi\rangle$ have the same statistics as those of $V|\psi\rangle$ for any initial state $|\psi\rangle$. Similarly, the operator norm is defined by:

$$||U|| = \sup_{|\psi\rangle \neq 0} \frac{||U|\psi\rangle||}{|||\psi\rangle||}.$$

It can be shown that the above distance measure is equivalent to the *trace distance* of the two unitary operators (up to normalization), which is easier to compute:

$$D_{\text{tr}}(U, V) = \frac{1}{2}||U - V||_1,$$

where $||X||_1 = \text{tr}(\sqrt{X^\dagger X})$ is the 1-norm of matrix X.

The unitary V is said to be an ϵ-*approximation* of the unitary operator U if $D(U, V) \leq \epsilon$. In the context of unitary synthesis, V is sometimes written as a sequence of instructions $g_1, g_2, \ldots, g_m$ from the gate set $\mathcal{G}$, that is, $V = g_m \cdots g_2 g_1$.

Errors Accumulate Linearly

A simple fact that enables many efficient approximate synthesis algorithms is the subadditivity of errors [9].

Lemma 6.6   *If $U_1, \ldots, U_t$ and $V_1, \ldots, V_t$ are operators such that $\|U_i\| \leq 1$, $\|V_i\| \leq 1$, and $\|U_i - V_i\| < \delta$, for all $i \in [t]$, then*

$$\|U_t \cdots U_2 U_1 - V_t \cdots V_2 V_1\| < t\delta.$$

This subadditivity property is the basis of many approximation algorithms.

Approximating Arbitrary Unitaries

Approximating an arbitrary unitary operator is generally very hard. Fortunately, any single-qubit gate can be approximated to arbitrary accuracy ϵ using generally $\Theta(1/\epsilon)$ gates from a finite gate set. A typical gate set of choice is the Clifford+T set, consisting of H gate, S gate, and T gate.

The Solovay–Kitaev (SK) algorithm shows that we can do much better—any single-qubit gate can be approximated to arbitrary accuracy ϵ_0 using only $O(\log^c(1/\epsilon_0))$ gates. The key idea for approximating a unitary U to arbitrary precision is to construct a finer and finer ϵ-net (i.e., a very small volume ball in $SU(d)$) around the identity using group commutator $VWV^\dagger W^\dagger$ repeatedly. Formally, the ϵ-ball can be defined as: $\mathcal{B}_\epsilon = \{U \in SU(d) : \|U - I\| \leq \epsilon\}$. Then the Solovay–Kitaev theorem states the following.

Theorem 6.7   *For any two universal instruction sets $S, T \subset SU(2^n)$ (closed under inverses, i.e., for any W in S or T, W^{-1} is also S or T) on n qubits, and a precision number $\delta > 0$, if a unitary transformation A consists of L instructions $U_1, \ldots, U_L \in S$, then A can also be implemented with precision δ using $M = O(L(\log(L/\delta)^c))$ instructions $V_1, \ldots, V_M \in T$ such that such that $\|U_L \cdots U_1 - V_M \cdots V_1\| \leq \delta$. (Here, c is between 3 and 4, and there is a classical algorithm for finding such circuit $V_i \in T$ in time $O(L(\log(L/\delta)^c))$).*

We now describe the SK algorithm for synthesizing a single-qubit gate U to arbitrary accuracy ϵ with a polylogarithmic running time in $1/\epsilon$ (Algorithm 6.1).

Algorithm 6.1 [220] is a recursive algorithm for approximating a unitary operator U with accuracy ϵ_n, where accuracy is better for larger n. Line 2 is the base case where a basic approximation to U is found by building a lookup table of length-ℓ sequences of gates, among which we pick the closest to U. In line 5, we find the appropriate group commutator such that $UU^\dagger = VWV^\dagger W^\dagger$. With this algorithm, it can be shown that accuracy improves as the recursive depth increases, that is $\epsilon_n < \epsilon_{n-1} < \cdots < \epsilon_0$. Details of the algorithm are omitted here; we refer the interested reader to the tutorial in [220].

An improvement to the base case search procedure has been developed by Amy et al. [221]. The "meet-in-the-middle" algorithm finds the minimal depth circuit with improved running

---

Algorithm 6.1 Solovay–Kitaev (SK) algorithm for synthesizing single-qubit gates

---

Function SolovayKitaev(single-qubit gate U, accuracy level n)

1: **if** $n == 0$ **then**
2:     **return** Closest approximation to U
3: **else**
4:     $U \leftarrow$ SolovayKitaev$(U, n-1)$
5:     $V, W \leftarrow$ Decompose $U U_{n-1}^{\dagger}$ into a balanced group commutator
6:     $V_{n-1} \leftarrow$ SolovayKitaev$(V, n-1)$
7:     $W_{n-1} \leftarrow$ SolovayKitaev$(W, n-1)$
8:     **return** $U_n \leftarrow V_{n-1} W_{n-1} V^{\dagger} W^{\dagger} U_{n-1}$
9: **end if**

---

time. The best known value for c is close to 2. It is still an exciting open problem to determine the best value for c (i.e., the shortest output sequence), possibly somewhere between 1 and 2.

Optimal Approximate Synthesis

We can bound the number of gates required for synthesizing (approximately) an arbitrary unitary transformation by a simple *volume counting* argument.

Recall in the Solovay–Kitaev theorem, we define the ϵ-ball in $SU(d)$ around a given unitary $V \in SU(d)$ as $B_{\epsilon} = \{U \in SU(d) : ||U - V|| \leq \epsilon\}$. It is understood that $SU(d)$ is a $d^2 - 1$ dimensional manifold, so the volume of the ball is proportional to ϵ^{d^2}. Hence, it takes $O((1/\epsilon)^{d^2})$ different strings of gates to represent the elements in $SU(d)$ to precision ϵ. Therefore, no algorithm can achieve synthesis of unitary in $SU(d)$ using fewer than $\log(1/\epsilon)$ gates.

A natural question to ask is: can some instruction sets saturate this lower bound? In other words, does there exist a instruction set S such that one can synthesize an arbitrary unitary in $SU(d)$ using $O(\log(1/\epsilon))$ gates? Fortunately, the answer is positive; at least for some universal instruction sets, such optimal synthesis algorithm exists. Harrow, Recht, and Chuang [222] showed a sufficient condition on S that allows for efficient universality.

6.1.4   HIGHER-DIMENSIONAL SYNTHESIS

Instead of using binary logic to target two-level qubits, compilers can target n-ary logic composed of qudits (i.e., d-level quantum systems). So far, in this book, we have been focusing on qubits alone; generalizations to high-dimensional logic has proven usefulness in circuit synthesis. In particular, quantum computations use a lot of temporary qubits (called ancilla). Ancilla bits are particularly necessary when performing arithmetic, since all quantum computations must be reversible (in order to conserve energy and avoid collapsing the quantum system to a classical state). Classical arithmetic can be computed with a universal NAND gate, but a NAND has two input bits and one output bit. It is impossible to reverse a NAND, since one output bit is not enough information to fully specify the two input bit values. The smallest universal reversible logic gate, which can simulate a NAND, has three inputs and three outputs, but one input and two outputs end up being ancilla. Consequently, arithmetic in quantum computations (which is very common), generates many ancilla. Qutrits (three-level systems) allow us to essentially generate ancilla by borrowing a third energy state in a quantum device to hold extra information.

Qudit synthesis can asymptotically change the critical path on quantum circuits. A number of recent algorithms [211, 223–226] have been developed for qutrit gate sets. In the NISQ era, the practical tradeoff, however, is that the third energy state generally is more susceptible to errors than the first two states used for binary qubits. Fortunately, noise models show that overall error can be decrease with qutrits. This is because computations require less time, and the third state is not always in use.

6.2   CLASSICAL VS. QUANTUM COMPILER OPTIMIZATION

Despite the unique properties of quantum circuits, we emphasize that quantum compiling is not completely different from classical compiling. As such, there are generally three classes of quantum compiling techniques:

- classical optimizations under classical constraints;

- classical optimizations under quantum constraints; and

- quantum optimizations under quantum constraints.

A quantum compiler must deal with control flow optimizations just like a classical compiler. The most notable *classical optimizations under classical constraints* that have been applied to quantum compiling include procedures such as loop unrolling, procedure cloning, interprocedural constant propagation, etc. The classical constraints, such as data dependencies, pipelining, and synchronizations, still apply in the context of quantum compiling. These optimizations inherited from classical compilation can significantly reduce the cost of a quantum circuit [170].

Fortunately, we have an abundance of techniques to inherit from classical compiler optimizations—for instance, the pass-driven approach in the LLVM compiler framework was adopted for quantum compiler.

IBM's Qiskit transpiler, as shown in Figure 6.5, is one of the examples of a pass-driven compilation framework. Multiple passes, each specialized in a different compilation task, are implemented in the framework. The user is free to chain together a subset of passes according to target program's needs.

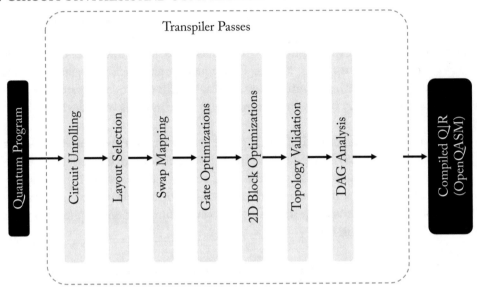

Figure 6.5: The compilation framework in Qiskit Terra developed by IBM [165].

Classical optimizations under quantum constraints play a crucial role in quantum compiling. This is where we need to adapt the familiar strategies in classical compilers to the unique quantum constrains. Communication cost of two-qubit interactions is one example of such quantum constraints. [161] has shown that routing strategies used in distributed systems can be borrowed after being adapted to the quantum communication cost. Register allocation algorithms also provide insights on how to allocate and reuse qubits in a quantum circuit [227]. Some other examples of such optimizations include mapping qubits to less noisy parts of a quantum machines [228–231], and scheduling parallel gates to avoid crosstalk between qubits [232], etc.

Finally, *quantum optimization under quantum constraints* refers to using quantum techniques to optimize for quantum compiling. For example, quantum programs can be stored in qubits (a quantum analogue of a stored-program architecture) using gate teleportation [233–235]. In this model, for example, quantum circuit synthesis tools that exploits the algebraic structures in unitary gates (see Section 6.1), and scheduling tools using teleportation [236], etc. Another example in this spectrum is to use a quantum computer to perform quantum compiling [237].

6.3 GATE SCHEDULING AND PARALLELISM

This section describes the impact of an important instruction-level optimization: gate scheduling. A *schedule* of a quantum program is a sequence of gate operations on logical qubits. The sequence ordering defines *data dependencies* between gates, where a gate g_2 depends on g_1 if they share a logical qubit and g_2 appears later in the schedule than g_1.

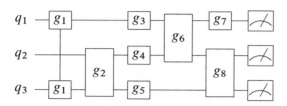

A quantum circuit executes from left to right. A qubit can only be involved in one quantum gate at a time. Data dependencies determine the sequential and parallel execution ordering of the gates in a quantum circuit. For example, two sequential gates back to back in general have strict ordering constraints:

$$q_1: \quad -\boxed{U}-\boxed{V}- \quad \neq \quad -\boxed{V}-\boxed{U}-$$

For two arbitrary unitary operators U and V, swapping the order of the two generally yields different results. One can quickly verify by writing down the matrix multiplications for two unitary matrices:

$$U \cdot V \neq V \cdot U.$$

Although not always equivalent, unitary matrices *can* sometimes be reordered in special cases. When that happens, we say the two matrices *commute* with each other. We will elaborate on the *commutation relations* later in the section.

Two parallel gates side by side in general has no ordering constraints:

$$q_1: \quad -\boxed{U}- \;=\; -\boxed{U}\!-\!- \;=\; -\boxed{U}-\boxed{I}- \;=\; --\boxed{U}-$$
$$q_2: \quad -\boxed{V}- \qquad --\boxed{V}- \qquad -\boxed{I}-\boxed{V}- \qquad -\boxed{V}--$$

Temporal ordering of parallel gates does not matter because we can check the corresponding tensor products and verify:

$$U \otimes V = (U \otimes I) \cdot (I \otimes V) = (I \otimes V) \cdot (U \otimes I).$$

The impact of gate scheduling can be quite significant in quantum circuits, and many algorithm implementations rely upon the execution of gates in parallel in order to achieve substantial

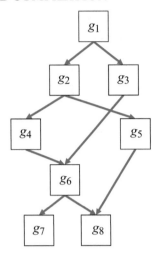

Figure 6.6: A data dependency graph for a quantum circuit with 8 gates: $g_1, \ldots, g_8$.

algorithmic speedup. Gate scheduling in quantum algorithms differs from classical instruction scheduling, as gate commutativity introduces another degree of freedom for schedulers to consider. Compared to the field of classical instruction scheduling, quantum gate scheduling has been relatively understudied, with only few systematic approaches being proposed that incorporate these new constraints.

6.3.1   PRIMARY CONSTRAINTS IN SCHEDULING

Data Dependency

Some gates are required to be applied after the completion of previous gates. This requirement can come from sharing of qubits, or specifications in the quantum algorithm (such as barriers and timing constraints). The previous section has illustrated the sequential and parallel ordering of quantum gates. It is convenient to construct a *data dependency graph* (DDG) when analyzing a quantum circuit. A DDG $G = (V, E)$ is defined as follows: each vertex v in V represents a quantum gate and a directed edge from v_1 to v_2, that is $(v_1, v_2) \in E$ means the gate in v_2 depends on the gate in v_2. Note that indirect dependencies are not drawn in the graph; if v_3 depends on v_2 and v_2 depends on v_1, we do not draw an edge for the transitive dependence between v_1 and v_3. For example, we can draw the DDG for a sample quantum circuit (omitting the measurements at the end), as shown in Figure 6.6.

Besides obeying the data dependency constraints to ensure logical correctness like a classical scheduler, a quantum scheduler must also respect other constraints such as the impact of noise while running a quantum circuit. In the following, we highlight a number of important considerations a quantum gate scheduler needs to take to ensure successful execution of a quantum circuit.

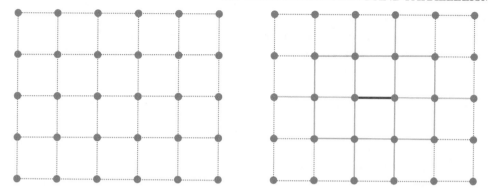

Figure 6.7: **Left:** the connectivity graph of a 5 × 6 2D mesh. Vertices are qubits, and two vertices are connected if the qubits are connected. **Right:** when the two center qubits are chosen for a two-qubit gate at an interaction frequency (highlighted in red), the surrounding qubits must be tuned off resonance from this interaction frequency, hence another two-qubit gate at the same frequency cannot be scheduled on any orange edges.

Hardware Limitations

To execute parallel instructions, a machine must be able to independently and simultaneously address and drive the individual qubits of interest. In NISQ computers, it usually comes with hardware constraints. For example, in a common 2D superconducting transmon architecture, each qubit is coupled at most with only its four nearest neighbors, which means that two-qubit gates can only be performed on adjacent qubits. Long-distance interactions are enabled by swapping qubits closer, hence inducing *communication overhead*. Furthermore, a NISQ machine typically has limited parallelism support. Although each qubit is connected independently to its drive line(s), sending pulse signals simultaneously down to two neighboring qubits could result in *crosstalk noise* depending on the coupling strength between them. In some design, simultaneous gates are prohibited if they are scheduled too close to each other, in order to prevent such crosstalk errors. Figure 6.7 illustrates the qubits that can be affected by crosstalk errors (to first order) if their frequencies were tuned improperly, assuming a frequency-tunable transmon architecture.

    A trapped-ion machine has unique scheduling constraints as well. In particular, the linear-trap tape-like architecture dictates that simultaneous gates can only be performed on contiguous qubits within a sliding window. This is due to the *hardware limitation* that qubits are controlled by laser beams of bounded width, typically smaller than the total number of qubits in a trap. One can only schedule a gate when (i) all previously dependent gates are scheduled and (ii) current laser beam window covers all operand qubit(s).

Commutation Relations

In a few cases, the sequential ordering constraints can be relaxed. As suggested in the last section, this can be illustrated by *commutation relations*. More specifically, we introduce the following definitions.

Definition 6.8   The *commutator* of two gates (operators) A, B is defined as

$$[A, B] = AB - BA.$$

If $[A, B] = 0$, then we say A and B commute.

In the context of scheduling, the commutation relation means that the two gates A, B are free to be reordered:

$$q_1 : \quad \boxed{A}\boxed{B} \quad = \quad \boxed{B}\boxed{A}$$

Finding two commuting operators may not be easy. A tool generalized from the commutation relation can be very useful when scheduling two sequential operators, namely, finding the conjugate of an operator.

Definition 6.9   Two gates A and B are *conjugate*, if there exists another gate U such that $UAU^\dagger = B$.

A simple rewrite of the above equality results in the following:

$$UAU^\dagger = B \implies UAU^\dagger U = BU \implies UA = BU.$$

As shown in the circuit model, we have found a way of moving one gate U to the right *through* another gate A (with the caveat that now A becomes B):

$$q_1 : \quad \boxed{U}\boxed{A} \quad = \quad \boxed{B}\boxed{U}$$

It is sometimes useful to construct the data dependency graph that reflects the commutation relations. So the *commutation-relaxed data dependency graph* (CRDDG) [238] is defined by first identifying groups of mutually commuting gates in the circuits, removing the dependencies of gates within a commuting set, and deriving new dependencies of any gates between two different sets. Suppose the gates in the example circuit can be grouped into the following

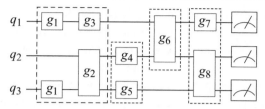

Figure 6.8: A sample quantum circuit with commutation relations indicated by dashed boxes.

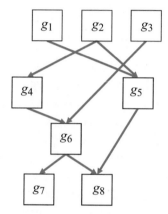

Figure 6.9: The commutation-relaxed data dependency graph (CRDDG) of the sample quantum circuit in Figure 6.8.

mutually commuting sets as shown in Figure 6.8. The corresponding CRDDG can be draw as in Figure 6.9.

Notice that the dependencies between g_1, g_2, and g_3 are removed due to the commutation relations. A less obvious additional dependency is from g_1 to g_5. Because g_5 would be directly adjacent to g_1, we could swap the order of g_1 and g_2 if allowed by their commutation relation.

Qubit Decoherence

In a NISQ machine, qubits have limited lifetime due to spontaneous decoherence. It is commonly referred to as idle noise on the qubits. A scheduler is responsible for maximize the utilization of the qubits within their lifetimes, that is from the moment they are initialized for computation until decoherence happens. Depending on the technologies, the number of gates safe to execute on a qubit before the chance of decoherence becomes too significant varies.

As a result, a common goal in scheduling quantum circuit is to minimize the *circuit depth*, that is the longest critical path of the circuit. It is closely related to the total *running time* of the circuit, and is sometimes called the total *latency* of the circuit. If the circuit depth is bounded within the lifetime of any qubits, then the circuit is generally safe from decoherence noise.

Another finer-grained measure that more precisely characterizes the utilization of qubits is the *space-time volume*. It accounts for the exact *active* usage (during which the qubits suffer from the risk of decoherence) of each qubit. Notice that when a qubit is in ground state $|0\rangle$, it is safe from decoherence. For instance, from the moment an ancilla qubit is freed (restored to ground state) until reused for computation, it stays safely in the ground state. For this reason, it does not contribute to the space-time volume. To accurately estimate the workload of a program, we define the space-time volume of a program as:

$$V = \sum_{q \in Q} \sum_{(t_i, t_f) \in T_q} \left(t_f - t_i \right),$$

where Q is the set of all qubits in the system, and T_q is a sequence of pairs $\{(t_i^0, t_f^0), (t_i^1, t_f^1), \ldots, (t_i^{|T_q|-1}, t_f^{|T_q|-1})\}$. Each pair corresponds to a qubit usage segment, that is we denote t_i^k and t_f^k as the allocation time and reclamation time of the kth time that qubit q is being used, respectively. V is high when a large number of qubits stay "live" (in-use) during the execution; thus, the higher the volume, the more costly it is to execute on that target machine.

One drawback of this definition is that it still does not fully reflect the impact of noise in the system. For instance, making the gate-dependent errors more explicit in the characterization will correlate the metric better with the cost for successful execution of the program. Space-time volume makes one step in that direction, but it remains an open problem to find a more effective metric.

6.3.2   SCHEDULING STRATEGIES

Some common scheduling strategies include the following.

- *ALAP (As-Late-As-Possible) Scheduling.* One of the simplest scheduling algorithms is the ALAP (As-Late-As-Possible) scheduler. In essence, it starts with the end of the circuit and schedules the last gates needed to be completed, and goes backward to their previous gates. The advantage of ALAP scheduler, as opposed to ASAP (as-soon-as-possible), is the qubits are initialized only when absolutely needed. Since qubits have limited lifetime, it is beneficial to initialize them as late as possible. This scheduer is implemented in Qiskit [165] by IBM.

- *LPF (Longest-Path-First) Scheduling.* When programs have more complex control flow, prioritization is needed between different parallel modules. The LPF (longest-path-first) scheduler tries to avoid increasing the critical path of the program, thus reducing the circuit depth. [160, 239] are some example LPF schedulers.

- *Communication-Aware Scheduling.* More advanced schedulers take into account costs of communication, due to two-qubit gates between operands that are far apart. Communication cost varies for different architectures, so does their scheduling strategies.

Here we highlight a general technique commonly used for reducing communication: *barrier insertion*. When a two-qubit gate is known to be causing high communication cost (such as long swap distance), separating it from other gates along the critical path can be effective. It is shown that iterative algorithms can benefit from the introduction of barriers as well, as inserting a barrier at the end of each round creates clean divisions between the rounds [161]. Details about communication cost can be found in Section 6.4 on qubit mapping and reuse.

- *Adaptive Scheduling*. When there are multiple implementations of the same gate, it is possible to let the scheduler choose the best one based on gate time, qubit fidelity, gate noises, etc. Due to our limited understanding of the noise characteristics of NISQ machines, this strategy remains a challenge.

Scheduling can be done at the logical level or the physical level. In NISQ machines, we just have the physical level as there is no error correction. While in error-corrected machines, we have both logical (fault-tolerant) and physical level. In the latter, it makes sense to apply these optimizations at both levels, that is to schedule the application program as well as to schedule the error correction routine [161, 240].

In the next section, we introduce another important compiler optimization step: qubit mapping and reuse. These optimizations are typically applied serially in a single pass in modern quantum compilers [165, 170], but it is not always clear which optimization should go first; sometimes it is necessary to redo the same optimization. The complexity goes up if one goes back to redo optimizations so as to obtain a better-optimized code.

6.3.3   HIGHLIGHT: GATE TELEPORTATION

In this section, we highlight a remarkable phenomenon called "quantum teleportation," and its important applications in gate scheduling, namely a strategy called "gate teleportation," in which scheduling gates is equivalent to scheduling resource states. We start by writing down an important resource state, the EPR pair $|\psi\rangle = \frac{|00\rangle + |11\rangle}{\sqrt{2}}$, which can be produced by the following circuit:

Here we introduce the fundamentals behind a teleportation-based quantum computer (QC) architecture, across which such EPR pairs are utilized as resource states. Because both quantum states and quantum gates can be transferred over long distance via a teleportation circuit, this technique is particularly useful in distributed QC architectures or in reducing communication cost in general.

Consider *scheduling a remote two-qubit gate* between two qubits that are too far to apply the joint gate pulse directly. In a non-teleportation-based QC architecture, one has to move the pair of qubits closer via physical movement or a series of swap gates, so as to perform the two-qubit gate. With teleportation, we have two more options, namely to move a quantum state to be closer to the other by teleporting the state, or to perform remote two-qubit gate by teleporting the gate.

Teleporting States

Suppose we have an arbitrary quantum state $|\psi\rangle$ and the EPR pair, the following circuit first destroys the state $|\psi\rangle$ via Bell-basis measurement, and subsequently restore $|\psi\rangle$ on one of the qubit in the EPR pair.

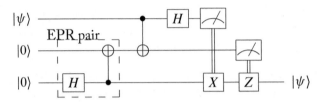

Notice that the boxed area is a EPR pair that can be prepared in advance. Now suppose the top two qubits are measured with outcomes $x, z \in \{0, 1\}$, respectively. Before the conditional X and conditional Z are applied, the bottom qubit is in a quantum state $|\phi\rangle = X^x Z^z |\psi\rangle$. Therefore, with appropriate recovery operations X and Z, we arrive at $|\psi\rangle$ in the bottom qubit.

This is called a "teleportation circuit," because one can interpret this procedure as follows: Alice holds the top two qubits ($|\psi\rangle$ and one of the qubits in the EPR pair prepared in advance), while Bob holds the bottom qubit. She wants to send him the quantum state $|\psi\rangle$. She can do so by performing the Bell measurements on her qubits, and send Bob two classical bits (e.g., over the phone). Upon applying the recovery gates X and Z, Bob can translate his qubit to $|\psi\rangle$.

Teleporting Two-Qubit Gates

Now consider an important application called "gate teleportation." It has been shown [233] that for certain unitary gate U, one can use the above circuit, to efficiently teleport a state "through" U so that the implementation cost and the communication cost for $U |\psi\rangle$ are significantly reduced. Specifically, we demonstrate the usefulness of a *remote CNOT gate* via teleportation. To begin with, we draw the following circuit:

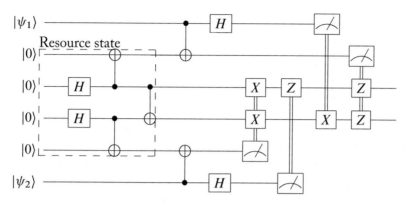

As a result, the two output qubits are $CNOT(|\psi_1\rangle, |\psi_2\rangle)$. Notice that the resource state $(|00\rangle + |11\rangle) |00\rangle /2 + (|01\rangle + |10\rangle) |11\rangle /2$ can be *prepared in advance*, because it does not depend on the input qubits $|\psi_1\rangle$ and $|\psi_2\rangle$. The rest of the circuit can be cut into two separate parts, i.e., first part consisting of the top three qubits and the second part consisting of the bottom three qubits. Notice that the two parts can be arbitrarily far apart. Hence, using this circuit, we have accomplished a remote CNOT gate, saving the cost of communication between the two qubits. We remark that the communication cost is essentially replaced by the preparation cost of the resource state; however, the resource state can be prepared offline (i.e., at compile time and prior to the execution of the circuit) and distributed across the systems in advance.

6.4   QUBIT MAPPING AND REUSE

Two convenient tools in analyzing qubit mapping are the *device connectivity graph* and the *program interaction graph*. A device connectivity graph, where each node is a qubit and two qubits are connected if the two qubits are allowed to directly interact. For example, in a superconducting device, this means whether or not two qubits are linked by circuit wires (through a coupler such as a capacitor); in a trapped ion device, this means whether or not laser beams can simultaneously address the two qubits. Connectivity graphs for superconducting device is commonly of the 2D mesh/lattice type, as shown in Figure 6.10. In contrast, trapped ion devices typically have much dense connectivity graphs, thanks to flexibility in performing two-qubit gates. Figure 6.11 shows some examples of trapped ion device connectivity graphs.

Given a quantum program, we can define a program interaction graph as a graph $G = (V, E)$ where V is a set of logical qubits present in the computation, and E is a set of two-qubit interaction gates contained in the program (e.g., CNOT gates). By analyzing this graph, we can perform an optimized *mapping*, which assigns a physical location for each logical qubit $q \in V$. The program interaction graph of the toy circuit can be constructed from enumerating all two-qubit gates, as shown in Figure 6.12.

Figure 6.10: Different connectivity graph for superconducting devices. **Left:** 2D square lattice. **Right:** 2D (heavy) hexagonal lattice. The choice of connectivity graphs is typically based on hardware constraints such as wiring bandwidth and noises of circuit components.

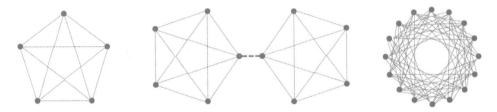

Figure 6.11: Different connectivity graph for trapped ion devices. **Left:** Complete (Clique) connectivity for small number of ions in one trap. **Center:** Weakly connected cliques for multiple traps. **Right:** A long chain of ions in one trap in a tape-like architecture.

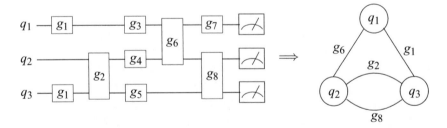

Figure 6.12: The program interaction graph for an example quantum circuit.

The goal of a mapper can be described as to *most efficiently embed a program interaction graph in a device connectivity graph*. Here efficiency means objectives such as minimizing communication and avoiding noisy qubits and links, etc. Once the program interaction graph is embedded, each edge (two-qubit interaction) is weighted by the cost of completing such interaction on the mapped qubits.

Cost of Long-Distance Interactions

In the context of NISQ machines, long-distance interactions are typically resolved by moving or swapping qubits. For simplicity, we call these as "routing qubits." Some architectures (such as trapped ion devices) support shuttling/transporting qubits directly, others (such as superconducting devices) bring two qubits together through a chain of swaps. Moving or swapping qubits not only costs time but also introduces errors.

Having the device connectivity graph is very convenient for analyzing these communication costs. We can model the overhead of long-distance interactions as weights on the edges of the connectivity graph. For example, the weighted distance from qubit q_i to qubit q_j is used to represent the cost of moving from location q_i to location q_j. Most existing mappers follow a two-step approach to find a good qubit mapping: (i) allocate qubits and (ii) route long-distance interactions.

Note that some work presents qubit routing as part of their scheduling strategies. But qubit mapping and gate scheduling are so intertwined that we expect them to be optimized together.

6.4.1   FINDING A GOOD QUBIT MAPPING

Three common heuristics which we analyze for *minimizing communication cost* (sometimes referred to as network congestion) are: edge distance minimization, edge density uniformity, and edge crossing minimization. In particular, the following analyses are performed on the program interaction graph mapped to the connectivity graph, denoted as the *mapped interaction graph*.

- *Edge Distance Minimization* The edge distance of the mapping can be defined as the Euclidean distance between the physical locations of each endpoint of each edge in the mapped interaction graph. Intuitively, in classical systems network latency correlates strongly with these distances, because longer edges require longer duration to execute. Similar situation applies to NISQ machines which communicate through swap chains. Longer-distance swap chains take more steps to complete and are more likely to overlap with other chains. It is worth noting that, for FT machines using surface code braiding operations [27], there is no direct correspondence between single braiding distance and time to complete the braid. However, longer surface code braids are more likely to overlap than shorter braids simply because they occupy larger area on the network, so minimizing the average braid length may reduce the induced network congestion as well [161].

- *Edge Density Uniformity* When two edges in the mapped interaction graph are very close to each other, they are more likely to intersect and cause congestion. Ideally, we would like to maximize the spacing between the edges and distribute them on the network as spread-out and uniformly as possible. This edge-edge repulsion heuristic therefore aims to maximize the spacing between operations across the machine.

- *Edge Crossings Minimization* We define an edge crossing in a mapping as two pairs of endpoints that intersect in their geodesic paths, once their endpoint qubits have been mapped. These crossings can indicate network congestion, as the simultaneous execution of two crossing operations could attempt to utilize the same resources on the network (e.g., swap through the same qubits). While the edge crossing metric is tightly correlated with routing congestion, minimizing it has been shown to be NP-hard and computationally expensive [241]. An edge crossing in a mapping also does not exactly correspond to induced network congestion, as more sophisticated routing algorithms can in some instances still perform these braids in parallel [242]. Some algorithms exist to produce crossing-free mappings of planar interaction graphs, although these typically pay a high area cost to do so [243].

With these objectives in mind, we demonstrate three procedures designed to optimize mappings, namely the graph partitioning approach, force-directed annealing approach, and community clustering approach.

Recursive Graph Partitioning

To compare against the local force-directed annealing approach, we also analyzed the performance of a global grid embedding technique based upon graph partitioning (GP) [244–246]. In particular, we utilized a recursive bi-sectioning technique that contracts vertices according to a heavy edge matching on the interaction graph, and makes a minimum cut on the contracted graph. This is followed by an expanding procedure in which the cut is adjusted to account for small discrepancies in the original coarsening [247, 248]. Each bisection made in the interaction graph is matched by a bisection made on the grid into which logical qubits are being mapped. The recursive procedure ultimately assigns nodes to partitions in the grid that correspond to partitions in the original interaction graph.

Force-Directed Annealing

The force-directed annealing [249–251] procedure consists of iteratively calculating cumulative forces and moving vertices according to these forces. Vertex-vertex attraction (to reduce edge length), edge-edge repulsion (to reduce edge density), and magnetic dipole edge rotation (to reduce edge crossing) are used to calculate a set of forces incident upon each vertex of the graph [161]. Once this is complete, the annealing procedure begins to move vertices through the mapping along a pathway directed by the net force calculation. A cost metric determines whether or not to complete a vertex move, as a function of the combination of average edge length, average edge spacing, and number of edge crossings. The algorithm iteratively calculates and transforms an input mapping according to these force calculations, until convergence in a local minimum occurs.

Community Clustering

In an interaction graph, subsets of qubits may interact more closely than others. These groups of qubits can be detected by performing *community detection analysis* on an interaction graph, including random walks, edge betweenness, spectral analysis of graph matrices, and others [252–257]. By detecting these structures, we can find embeddings that preserve locality for qubits that are members of the same community, thereby reducing the average edge distance of the mapping and localizing the congestion caused by subsets of the qubits.

To respect the proximity of the vertices in a detected community, we break up our procedure into two parts: first, impose a repulsion force between two communities such that they do not intersect and are well separated spatially; and second, if one community has been broken up into individual components/clusters, we join the clusters by exerting attracting forces on the clusters. In particular, we use the KMeans clustering algorithm [258, 259] to pinpoint the centroid of each cluster within a community and use them determine the scale of attraction force for joining them.

6.4.2   STRATEGICALLY REUSING QUBITS

Reclaiming Qubits via Measurement and Reset

When some qubits are disentangled from the data qubits, we can directly reclaim those qubits by performing a measurement and reset. We can save the number of qubits, by moving measurements to as early as possible in the program, so early that we can reuse the same qubits after measurement for other computation. Prior art [260, 261] has extensively studied this problem and proposed algorithms for discovering such opportunities.

This measurement-and-reset (M&R) approach has limitations. First, today's NISQ hardware does not yet support fast qubit reset, so reusing qubits after measurement could be costly or, in many cases, unfeasible. The state-of-the-art technique for resetting a qubit on a NISQ architecture is by waiting long enough for qubit decoherence to happen naturally, typically on the order of milliseconds for superconducting machines [51], significantly longer than the average gate time around several nanoseconds. FT architectures have much lower measurement overhead (that is roughly the same as that of a single gate operation), and thus are more amenable to the M&R approach. Second, qubit rewiring as introduced in [261] works only if measurements can be done early in a program, which may be rare in quantum algorithms—measurements are absent in many program (such as arithmetic subroutines) or only present somewhere deep in the circuit. Unlike the uncomputation approach, M&R does not actively create qubit reuse opportunities.

Reusing Qubits via Borrowing

Another strategy for reusing qubits involve temporarily borrowing a qubit for computation and return the qubit to its original state when completed. This technique is sometimes called the "dirty borrowing" of qubits, because the qubits we borrow can be in an arbitrary unknown quan-

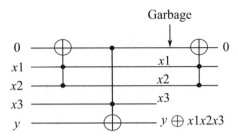

Figure 6.13: Implementation of $\Lambda^3(X)$ gate using an ancilla qubit in $|0\rangle$.

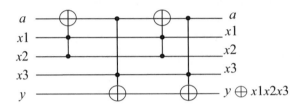

Figure 6.14: Implementation of $\Lambda^3(X)$ gate using an ancilla qubit in an arbitrary state.

tum state; this is to be contrasted with the uncomputation technique we will introduce next, in which the qubits to reuse are always clean ancilla (i.e., qubits initialized to a known state such as $|0\rangle$).

For example, Figures 6.13 and 6.14 demonstrate the implementation of multi-controlled NOT gate such as $\Lambda^3(X)$ via borrowing of clean and dirty ancilla qubits, respectively.

Dirty borrowing opportunities depends highly on the structures in quantum circuits; the reason is two-fold. First, we need to return the borrowed qubits to their original states timely, otherwise the original computation has to be stalled. Second, the borrowing computation is restricted. For example, one typically cannot perform measurements on the borrowed qubits, due to entanglement with other qubits. Because the quantum state of the dirty ancilla is unknown, the borrowing circuit typically rely on the difference between $|\psi\rangle$ and $X|\psi\rangle$, instead of the state $|\psi\rangle$ itself, to perform computation. This technique has been applied to speed up implementations of arithmetic circuits as found in [262, 263].

6.4.3   HIGHLIGHT: UNCOMPUTATION

Reclaiming ancilla qubits that are entangled with data qubits is non-trivial, as measuring and resetting them will alter the data qubits' state. Fortunately, *uncomputation*, introduced by Bennett [74], is the process for undoing a computation in order to remove the entanglement relationship between ancilla qubits and data qubits from previous computations. Figure 6.15 illustrates this process. In that circuit diagram, the U_f box denotes the circuit that computes a classical function f. The garbage produced at the end of U_f is cleaned up by storing the out-

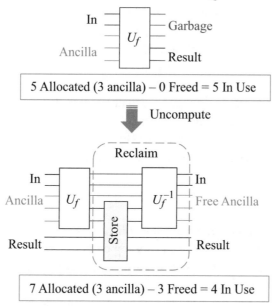

Figure 6.15: Ancilla qubit reclamation via uncomputation. Each horizontal line is a qubit. Each solid box contains reversible gates. Qubits are highlighted red for the duration of being garbage.

put elsewhere and then undoing the computation. This "storing" step can be done by applying qubit-wise CNOT gates controlling on the output qubits and targeting on the copy qubits. One could alternatively accumulate the output values to some result qubits, using operations such as in-place addition [212]. The "uncomputation" step is accomplished by applying all gates in U_f in reverse order, denoted as U_f^{-1} in the diagram. In the end, all output qubits are safely stored and all ancilla qubits are reset to their initial value, typically in the $|0\rangle$ states.

This uncomputation approach has two potential limitations: first, we need to pay for the additional gate cost, and second, it only works if the circuit U_f is a classical reversible transformation—i.e., can be implemented with Toffoli gate alone, optionally with NOT gate and CNOT gate. This excludes the common quantum gates such as H gate and Phase gate. Nonetheless, classical reversible computation implements the arithmetic logic on a quantum computer, which plays a large part in many quantum algorithms [11, 13, 214].

The optimization of qubit allocation and reclamation in reversible programs dates back to work as early as [213, 264], where they propose to reduce qubit cost via fine-grained un-computation at the expense of increasing time. Since then, more work [265–268] follows in characterizing the complexity of reclamation for programs with complex modular structures. Recent work in [201, 212] show that knowing the structure of the operations in U_f can also help identify bits that may be eligible for cleanup early. A more recent example [269] improves the reclamation strategy for straight-line programs using a SAT-based algorithm. Some of the above work emphasizes on identifying reclamation opportunities in a flat program, whereas our focus is on coordinating multiple reclamation points in a larger modular program.

Most existing qubit reuse algorithms using uncomputation [212, 213, 264] emphasize on the asymptotic qubit savings, and commonly make an ideal assumption that machines have all-to-all qubit connectivity (i.e., no locality constraint). Since all qubits are considered identical, a straightforward way to keep track of qubits is to maintain a global pool, sometimes referred to as the *ancilla heap*. Ancilla qubits are pushed to the heap when they are reclaimed, and popped off when they are allocated, for instance in a last-in-first-out (LIFO) manner. In this ideal model, we can simply track qubit usage by counting the total number of fresh qubits ever allocated during the lifetime of a program.

Reversible Pebbling Game

Uncomputation exposes a very interesting tradeoff between space and time costs of a circuit. One may worry the substantial gate cost in running computation forward and backward is too high to be practical. Fortunately, it is possible to significantly reduce space by clean up garbarge qubits by strategically uncomputing parts of the circuit during the course of computation. The problem resembles a game, i.e., the *reversible pebble game*. Suppose we divide a computation into m sequential steps of roughly equal size, each of which is represented by a node in a graph. The dependency graph thus looks like a length-m directed path. Consider executing step k as placing a pebble on the kth node of the graph, and reversing step k as removing a pebble from the node. The goal of the game is to place a pebble on the mth node, under the rule that we can place or remove a pebble on node k if $k = 1$ or node $k - 1$ has a pebble.

We now describe a divide-and-conquer strategy that reaches node m using only $\log m + 1$ pebbles. Suppose we divide a computation into $m = 2^d$ steps of roughly equally size. Then we can construct a program call graph that is a *balanced binary tree* of depth d, where the leaves contain the steps in the computation, and all internal nodes have two children function calls. We execute the program by traversing the program call graph in depth-first search order. In particular, when we enter a node in the tree, we execute the corresponding function forward; when we exit a node in the tree, if it is the left child of its parent, we execute the function backward. Pictorially, we demonstrate the process for $d = 2$ in Figure 6.16.

As a result, we follow a recursive pattern of two steps forward and one step backward as we proceed in the computation. That is, for every node f_P whose two children are denoted as f_L and f_R:

$$f_P = f_L f_R f_L^{-1}.$$

Assume each step in the computation takes unit time. We found that the total time to execute the program without uncomputation is $T = 2^d$. In comparison, the total time to execute the program with the divide-and-conquer approach is $T' = 3^d \sim T^{\log_2(3)} = T^{1.585}$. In principle, one can further reduce the power to close to 1 by dividing computation into ℓ^d steps and constructing a balanced ℓ-ary graph where the first $\ell - 1$ children are uncomputed.

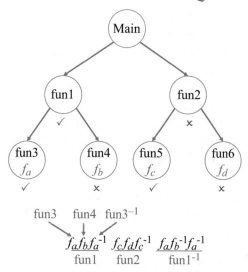

Figure 6.16: Divide-and-conquer approach to reversible pebble game. The balanced binary tree is traversed in depth-first search order. Function is executed when entering a node in the tree; function is uncomputed when exiting a node that is a left child of its parent (denoted by blue check marks).

In a realistic setting, the gate cost and qubit cost for each step can be different. One important factor to consider is the locality of qubits in a quantum architecture. A two-qubit gate, such as a CNOT gate, takes time proportional to the communication cost between the two operand qubits. So, in general, uncomputation needs to be performed frequent enough to prevent garbage from accumulating, making available qubits with good locality for reuse. In the next section, we demonstrate how to use a heuristic-based uncomputation algorithm to integrate qubit allocation with reclamation under realistic architectural constraints.

Heuristic-Based Noise-Aware Uncomputation
To more realistically estimate the cost of running algorithms on a machine with qubit locality constraint, we need to take the communication overhead into account [227]. Suppose we are given a program with n functions that potentially have qubit reclamation. Our goal is to determine the optimal choices of whether to reclaim at each of those n locations.

Qubit allocation can benefit from locality awareness. A good algorithm prioritizes qubits according to their locations in the machine. At a high level, it chooses qubits from the ancilla heap by balancing three main considerations—communication, serialization, and area expansion.

When deciding which qubits to allocate and reuse, one approach is to use a heuristic-based algorithm that assigns priorities to all qubits. The priorities are weighted not only by the communication overhead of two-qubit interactions but also by their potential impact to the parallelism of the program. Reusing qubits adds data dependencies to a program and thus serializes computation, but not reusing qubits expands the area of active qubits and thus increases the communication overhead between them.

Uncomputation decisions can be made with a simple *cost-benefit analysis*: at each potential reclamation point, we estimate and compare two quantities: (i) cost of uncomputation and reclaiming ancilla qubits; and (ii) cost of no uncomputation and leaving garbage qubits. To do so, we need an efficient way to accurately estimate the C_1 and C_0 quantities. This is a non-trivial task. In particular, the decision of child function affects not only the cost of itself, but also the cost of parent function. If a child function decides to uncompute, the additional gate costs need to be duplicated should its parent decide to uncompute as well. This is a phenomenon called "recomputation." Thus, we should take the level of the function into account when we make the decision.

6.5   SUMMARY AND OUTLOOK

A generic tool flow for quantum programming and compilation includes multiple layers of transformations and optimizations. A quantum algorithm is implemented in a quantum program using a quantum domain specific language (DSL) (see Chapter 5). The program is then translated into quantum intermediate representation (QIR), undergoing a series of transformation in the compiler, including circuit synthesis and optimization, gate scheduling, and qubit mapping, etc. The QIR is then translated into analog pulse sequences for control qubits (see Chapter 7). [31, 270] discuss the design of quantum computer architectures in greater details.

The most widely used approach for combining all transformations in the compiler is called the pass-driven approaches, where each transformation/optimization is applied once in sequence. There is another approach called instrument-driven (commonly used in LLVM) is somewhat strange in that we produce an classical executable program that generates the output QIR. For instance, we take a C program with quantum code, turn into an executable classical program which has all the classical control such as conditionals and loops and then print out the program as it runs. One of the advantages to this approach is to make the analysis more tractable. With really large programs and really large machines, the pass-driven approach would occupy huge amount of memory. This is more scalable. For NISQ machines, scalability is usually not our main concern, making it optimal is more important.

Further Reading

Here we highlight some other exciting research directions in software support for NISQ computers. Efficient logic synthesis methods for higher-dimensional quantum systems are not so well understood. It remains an open problem to produce optimal circuits for logic in d-level systems (with qudits, as opposed to qubits) [226, 271–274]. Even though we have divided the discussions on quantum computer systems as different layers. They are intertwined. In fact, cross-layer optimizations are believed to be critical for bridging the gap between algorithms and hardware [31, 275]. The growing complexity in compiler transformations also necessitates the verification of the compiler.

Quantum circuits need to be highly optimized so as to fit in the exacting resource constraints. Hence, an active research direction is the logical (instruction-level) optimizations of quantum circuits. A wide range of techniques have been developed to optimize various aspects of quantum circuits. For example, circuit template optimization recognizes circuit patterns and substitutes them with more efficient substitutions. In a nutshell, given a start state and an end state, we aim to find the shortest path between those two states. This is to be differentiated with optimizations using pulse compilation, where we look for the most efficient physical pulse to implement a native instruction as accurately as possible. In this case, at the compiler level, we substitute a sequence of instructions that is logically equivalent to the ideal sequence. With that being said, circuit optimization and pulse optimization are not without connections. For instance, sometimes a more accurate pulse can be implemented using one sequence of instructions than the other, so knowledge of pulse control can sometimes inform a compiler which compiled program is preferable.

CHAPTER 7

Microarchitecture and Pulse Compilation

One cannot really talk about building scalable quantum computer systems without bringing the discussion of an architectural support for the increasingly complex control and memory modules to the table. As we enter the NISQ era and beyond, we will need to orchestrate the simultaneous quantum operations on hundreds or thousands of qubits. What kind of microarchitecture can keep up with the speed and bandwidth of quantum information processing? How do we build a reliable interface between classical control/feedback signals and quantum data? Can we efficiently translate and synchronize machine pulses from gate instructions? This may be feasible by hand for small-scale devices with 5–10 qubits, but it will soon become intractable without an automated, robust control system as the size of the devices scales up.

In this chapter, we pay particular attention to three aspects of such systems: classical and quantum control of qubits, pulse generation and optimization, and calibration and verification. "From Gates to Pulses" describes the general flow for constructing pulse sequences. "Quantum Controls and Pulse Shaping" illustrates the progress and challenges in managing the classical and quantum datapath required for a large number of qubits under tight time, power and bandwidth budgets. Next, "Quantum Optimal Control" shows the general principles in translating quantum gates to hardware pulses, and demonstrates two novel techniques, each targeting higher pulse quality and faster compilation time.

7.1 FROM GATES TO PULSES–AN OVERVIEW

7.1.1 GENERAL PULSE COMPILATION FLOW

At the lowest level of control hardware, qubits are driven by analog pulses. Recall from Chapter 2 that, depending on the types of the qubits, these pulses are sent in different forms, e.g., modulated lasers for trapped ion qubits, and microwave electric signals for superconducting transmon qubits. Therefore, quantum compilation must translate from a device-independent high-level quantum program down to a sequence of device-dependent control pulses.

Figure 7.1 shows the general flow for compiling analog pulses. The input to this process is a sequence of quantum instructions (produced by logical-level compilations and optimizations including scheduling and mapping), and the output is a sequence of analog pulses that implements the logical instructions.

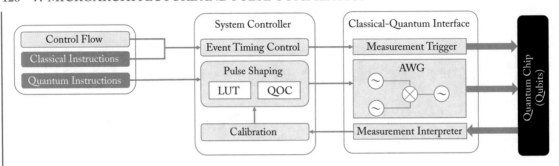

Figure 7.1: General flow for translating quantum gates to analog pulses.

The simplest approach to perform such gate-to-pulse translation is by using a lookup table (LUT). Once a quantum algorithm has been decomposed into a quantum circuit comprising single- and two- qubit gates, the compiler simply proceeds by concatenating a sequence of pulses corresponding to each gate. In particular, a lookup table maps from each gate in the gate set to a sequence of control pulses that executes that gate.

For example, in a superconducting architecture, the $\{R_x(\theta), R_z(\phi), CX\}$ gate set alone is sufficient for universality, so in principle the H and SWAP gates could be removed from the compilation basis gate set. However, we include the generated pulses.

The advantage of the LUT approach is its short pulse compilation time, as the lookup and concatenation of pulses can be accomplished almost instantaneously. However, it prevents the optimization of pulses from happening across the gates, because there might exist a global pulse for the entire circuit that is shorter and more accurate than the concatenated one. The quality of the concatenated pulse relies heavily on an efficient gate decomposition of the quantum algorithm.

Another generic technique for generating pulses is through pulse shaping tools based on quantum optimal control (QOC) theory, which is illustrated in detail in Section 7.3.

The next section delves into pulse shaping techniques and introduces the design principles of quantum controls in greater details.

7.2   QUANTUM CONTROLS AND PULSE SHAPING

Controlability of a quantum system is very much a fundamental issue. The aim is to establish a framework of strong theoretical understanding and practical methodology for driving a physical system to a desired state. This is challenging because quantum systems exhibit unique characteristics (such as coherence, superposition, and entanglement). One of the most prominent distinctions between classical control and quantum control is the difficulty in acquiring information about the state of a quantum system without disturbing it, which makes feedback

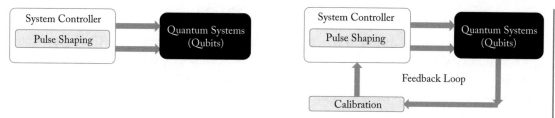

Figure 7.2: Open-loop control (left) vs. closed-loop control for pulse shaping.

control non-trivial. In the following, we describe some leading strategies for qubit controls and demonstrate the challenges in crossing the classical-quantum boundary.

7.2.1   OPEN-LOOP VS. CLOSED-LOOP CONTROL

Quantum control techniques can be generally classified into two categories: (i) open-loop control and (ii) closed-loop control. The primary distinction between the two is whether there is real-time feedback from the quantum systems—in open-loop control, systems output is not continuously monitored, whereas in closed-loop control, the controller calibrates the pulses based on continuous feedback. Figure 7.2 pictorially illustrates the distinction.

Studies in open-loop control generally fall into three categories. First, some work focuses on the optimality and reachability of pulses for different quantum systems (i.e., Hamiltonians). One widely used approach is to express the controllability criteria in terms of structures in Lie groups and Lie algebras [276], or in terms of graph theoretical concepts [277, 278]. Second, numerical optimal control theory has shown to produce versatile and realistic pulse sequences. It uses numerical methods to search for the best way of achieving given quantum objectives in shortest time and most realistic shape. This strategy is sometimes referred to as open loop coherent control [279–281]. Third, Lyapunov-based control design [282, 283] is another useful approach in open-loop quantum control. In this approach, the control input is determined by the state of the systems. In quantum control, however, it is non-trivial to obtain information about quantum states without disturbing them. An open-loop control design is used to simulate an artificial closed-loop system. Closed-loop control is believed to be more robust and reliable. There are in general two different approaches in closed-loop control [284]: (i) learning control and (ii) feedback control. In closed-loop learning control, each cycle of the loop is executed on a new copy of the system [285]; whereas in a feedback control, information about the states of the quantum systems is continuously fed back to the controller through measurement or state estimation [286].

For more details on quantum control, we refer the interested readers to tutorials [276, 287–289].

7.3 QUANTUM OPTIMAL CONTROL

Given a general setting of quantum systems, GRadient Pulse Engineering (GRAPE) is a strategy for numerically finding the best control pulses for computation by following a gradient descent procedure [290, 291]. This is sometimes viewed as the last step in quantum circuit synthesis—the output of GRAPE is the control pulse parameters needed for the underlying hardware architecture.

The basics of GRAPE can be illustrated by examining the Hamiltonian picture of the physical systems. Consider a quantum system with intrinsic Hamiltonian $\mathcal{H}_0$ and a set of external controls, in the form of time-dependent Hamiltonian operators, $\{\mathcal{H}_1, \mathcal{H}_2, \ldots, \mathcal{H}_m\}$. The overall system Hamiltonian can be written as

$$\mathcal{H}(t) = \mathcal{H}_I + \sum_{i=1}^{m} \mathcal{H}_i(t),$$

where $\mathcal{H}_i(t)$ is typically a product of time-dependent dimensionless amplitude and a time-independent control operator. That is,

$$\mathcal{H}_i(t) = u_i(t)\mathcal{H}_i).$$

The objective of the optimal control theory is to find the control fields $\{u_1(t), u_2(t), \ldots, u_m(t)\}$, such that the $\mathcal{H}(t)$ approximates the target unitary $\hat{U}$ to high precision.

What is the implemented unitary as a result of the Hamiltonian evolution? Solving for the solution to a continuous time-dependent equation is challenging, so we discretize the time interval into sufficiently small time steps δt. This is called the *piecewise-constant approximation*. Suppose we evolve the quantum systems, from t_0 to the jth timestep, $t = t_0 + j\delta t$. For timestep j, the set of *constant* control fields is denoted by $\{u_{1,j}, u_{2,j}, \ldots, u_{m,j}\}$. Then the time-independent Hamiltonian for timestep j is

$$H_j = \mathcal{H}_0 + \sum_{i=1}^{m} u_{i,j}\mathcal{H}_m.$$

Therefore, the unitary operation accomplished at timestep j is

$$U_j = e^{-iH_j\delta t}.$$

Evolving from t_0 to t_j by the piecewise-constant approximation can be written as

$$U = U_j U_{j-1} \cdots U_2 U_1.$$

GRAPE performs gradient descent search over the space of possible control field parameters that approximate the targeted unitary matrix up to a specified fidelity. In general, if the input state to the computation is known, GRAPE can optimize for a control pulse that works for that particular input state. In other words, we can find an approximation $\hat{U} \neq U$, but

$$|\psi_{out}\rangle = \hat{U} \, |\psi_{in}\rangle \approx U \, |\psi_{in}\rangle \, .$$

Besides the final fidelity, the set of control field parameters must satisfy some constraints, such that the resulting control pulses are physically realizable. Following the analysis in [292], we name a few such constraints.

- The amplitude of each control field $|u_j|$ parameter is to be minimized.

- Each control parameter needs to form a smooth pulse over time; $\sum_{i,j} |u_{i,j} - u_{i,j+1}|$ is to be minimized.

- Evolution time, i.e., pulse time, is to be minimized. Because qubits have short lifetimes due to quantum decoherence.

Minimizing the cost function is the key in quantum optimal control. To be able to apply GRAPE, the cost functions need to be differentiable. The gradients are computed analytically and backpropagated with automatic differentiation. GRAPE has been used to mitigate other sources of error such as gate errors, State Preparation and Measurement (SPAM) errors, and qubit crosstalk, as demonstrated in past work [293–295].

It remains open problem to speedup the GRAPE algorithm, as state-of-the-art techniques scale up poorly. While GRAPE on one- or two-qubit unitaries might find the optimal pulse sequence quickly, it would take an unreasonable amount of time for circuits with five or more qubits. In the following, we will see a systems approach to speedup the pulse compilation process for variational algorithms.

7.3.1   HIGHLIGHT: COMPILATION FOR VARIATIONAL ALGORITHMS

The conventional approach for compiling large programs is to synthesize a quantum program using a small set of primitive gates, apply quantum optimal control theory to each primitive gates, and string together all pulses to accomplish the computation. While this approach results in a constant time compilation (as we can build a lookup table for the pulses of the primitive gates and reuse them for every circuit), it does not produce the optimal pulse sequence (e.g., shortest pulse length). On the other spectrum, while full quantum optimal control generates the fastest possible pulse sequence for a target circuit, its compilation latency is large due to time spent on GRAPE.

As such, *partial compilation* has been proposed [159], which compiles only parts of a quantum circuit using GRAPE while leaving the rest in a lookup table, balancing optimality in pulse length and pulse compilation time.

This approach is particularly useful for variational quantum algorithms. Here the quantum circuit are parametric as seen in Chapter 3, meaning that there are some parts of the circuit that are fixed, and there are other parts that change according to a set of parameters. In a variational algorithm, every iteration of quantum circuits differs only slightly. Hence, we can use GRAPE to recompile only the changing parts, yielding lower cost than recompiling the entire circuit.

7.4   SUMMARY AND OUTLOOK

Further Reading

Quantum control theory has already attained significant success since the beginning of quantum computation research [284, 296, 297]. But scalability (in compute time and memory) of the pulse optimizer is still an issue. More intelligent methods must be developed for high-dimensional space optimizations, as gradient techniques generally do not work well in higher-dimensional spaces. Developing methods for modifying pulses remains an exciting open area. This is a well-motivated problem because it is potentially beneficial to leverage calibrated pulses for different machines, to use one pulse as a guide for optimization on another, or to combine pulses for subsystems into one composite pulse [298]. Techniques such as dynamical decoupling, are extremely effective in error mitigation by interrupting quantum systems by π-pulses [299–302]. Recent efforts have used quantum optimal control theory to speed up pulses for a range of quantum algorithms [159, 292, 303–305].

Controlling quantum computation is difficult, in part because of the boundaries between the classical and quantum world. One such example is the scale boundary—there is a discrepancy in scale between the qubit object and the classical control mechanism, e.g., [131, 306] raise the issue of pitch-matching. Another example is the environment boundary—in order for qubits to behave quantum mechanically, they may need to be put in some critical environment, e.g., vacuum, and low temperature. As a result, the control mechanism will have to cross those boundaries so as to send the signals to the right qubits with high precision. Some have tried integrating classical control circuits into the cryogenic systems of their quantum device [307–309] to reduce the bandwidth of the classical control signal needed for keeping up with the computation.

CHAPTER 8

Noise Mitigation and Error Correction

For a quantum computer to work, it takes extreme precision to isolate and control qubits. In particular, qubits are very short-lived due to the interactions with the environment. Quantum logic gates can have small drifts when pulses are out-of-focus or out-of-tune. Classical controls and calibrations may have a hard time keeping up in scale, speed, and power. If there were no strategies to overcome the aforementioned examples of noises, they would accumulate and eventually lead to critical failure in computation. Physical noises in current devices put stringent limitations on their computing capabilities. Indeed, this chapter aims to address the central issue: how do we protect information from the adverse impacts of noise? We highlight two classes of strategies that have been developed for information protection and error reduction in quantum systems, namely noise mitigation and error correction. In both cases, we demonstrate a few promising examples that effectively lower the error rates, and then stress the drawbacks and challenges that still remain.

8.1   CHARACTERIZING REALISTIC NOISES

From an engineer's perspective, the making of a practical quantum device is a process where we iteratively improve our capability of estimating and fixing sources of errors. The typical engineering cycle is as follows: as we understand the sources better, we change the focus to the next dominant errors and try to measure and fix them again, until we reach some point that many hope to achieve, namely a fault-tolerant quantum computer. As of now, in the NISQ era, our experience with quantum computing devices is still nascent.

The state-of-the-art tools we have at hand for measuring sources of errors include (i) state and process tomography and (ii) randomized benchmarking. The goal of these tools are to efficiently characterize noise sources.

First we need to understand how to quantitatively study noise. The effects of noise on a quantum system are typically probabilistic. Starting with a (pure) quantum state, a quantum system undergoes a spontaneous interaction with the environment, which results in a probability distribution of quantum states. As such, we revisit our definition of mixed quantum state in Chapter 2.

Definition 8.1   A mixed quantum state can be written as a probability distribution over some pure quantum states:

$$\rho = \sum_i p_i \, |\psi_i\rangle \, \langle\psi_i| \, .$$

ρ is called the density matrix of the quantum state. Naturally, we care about the "distance" between results obtained in two scenarios: ideal and noisy. Two common measures of distance between two states are *fidelity* and *trace distance*.

Definition 8.2   The fidelity between two states is defined as

$$F(\rho, \sigma) = ||\sqrt{\rho}\sqrt{\sigma}||_1^2 \, .$$

Definition 8.3   The trace distance between two states is defined as

$$D_{\text{tr}}(\rho, \sigma) = \frac{1}{2}||\rho - \sigma||_1 \, .$$

Notice that both measures are bounded between 0 and 1, and are symmetric. In fact, fidelity is related to the trace distance as follows:

$$1 - \sqrt{F(\rho, \sigma)} \leq D_{\text{tr}}(\rho, \sigma) \leq \sqrt{1 - F(\rho, \sigma)} \, .$$

Suppose we want to evolve a quantum system with some process $\mathcal{U}$. However, in reality, we end up applying a noisy version of the process $\mathcal{U}$. We introduce two common ways of quantifying the noise of a process, namely *average error rate* and *diamond distance*. We write the effect of noise on a quantum state as a *channel*: $\rho \to \mathcal{E}(\rho)$.

Definition 8.4   The average error rate of a process $\mathcal{U}$ under a noise channel $\mathcal{E}$ can be written as:

$$r(\mathcal{U}, \mathcal{E}) = 1 - \int d\psi \, \langle\psi|U^\dagger \mathcal{E}(\psi) U |\psi\rangle \, .$$

The underlying physical interpretation of this measure is that we send a pure quantum state $|\psi\rangle$ through a noisy evolution and back, compute what is the probability of getting the initial state back, and then average over all possible pure states uniformly and randomly. The state is bounded between 0 and $d/(d + 1)$, where d is the dimension of the Hilbert space.

Another measure of noise is the diamond distance between two processes.

Definition 8.5   The Diamond distance (completely bounded norm) of two processes $\mathcal{P}$ and $\mathcal{Q}$ is defined as:

$$D(\mathcal{P}, \mathcal{Q}) = \sup_{\rho} \frac{1}{2} ||[\mathcal{P} \otimes \mathcal{I} - \mathcal{Q} \otimes \mathcal{I}](\rho)||_1.$$

The supremum is taken over all possible mixed quantum states. The interpretation of diamond distance is that it measures the *worst case* difference between two channels based on single-shot measurements.

The strategies for characterizing noise are quantum state/process tomography and randomized benchmarking. The former reveals full information about a quantum state or process via complex measurements, while the latter reveals partial information through efficient procedures.

8.1.1   MEASUREMENTS OF DECOHERENCE

As shown in Section 2.3.3, two canonical measures of the robustness of a qubit are the T_1 relaxation time and the T_2 relaxation time. In this section, we briefly discuss how to experimentally determine the T_1 and T_2 decoherence rates.

Measuring T_1

To characterize T_1 relaxation, the standard approach is to quantify the rate of exponential decay using simple steps as follows.

1. Initialize qubit to ground state $\rho = |0\rangle \langle 0|$.

2. Apply X gate, i.e., $\begin{pmatrix} 0 & 1 \\ 1 & 0 \end{pmatrix}$ to the qubit, so $\rho' = X\rho X$.

3. Wait for time t arriving at ρ''.

4. Measure the probability of ρ'' being in $|1\rangle \langle 1|$.

Measuring T_2

The standard approach to quantify the rate of exponential T_2 decay is as follows.

1. Initialize qubit to ground state $\rho = |0\rangle \langle 0|$.

2. Apply H gate, i.e., $\frac{1}{\sqrt{2}} \begin{pmatrix} 1 & 1 \\ 1 & -1 \end{pmatrix}$ to the qubit, so $\rho' = H\rho H$.

3. Wait for time t arriving at ρ''.

4. Apply H gate again, so $\rho''' = H\rho'' H$.

5. Measure the probability of ρ''' being in $|0\rangle \langle 0|$.

8.1.2   QUANTUM-STATE TOMOGRAPHY

The goal of quantum state tomography [310, 311] is to reconstruct an unknown quantum state ρ based on the outcomes from a series of measurements. Suppose we define E_i as the projector for a particular measurement outcome e_i, then the probability of obtaining this outcome when measuring ρ can be written as $\mathbf{Pr}[e_i|\rho] = \text{tr}(E_i\rho)$ by Born's rule. For all possible measurement outcomes, we can construct a histogram of observations for each measurement. It turns out this sampling process is linear, that is $A\rho = p$, where p is the probabilities of measurement outcomes. Then we can use *linear inversion* to reconstruct ρ from p, i.e., $\rho = (A^T A)^{-1} A^T p$. Experimentalists commonly use the *maximum likelihood estimator* (instead of linear inversion) to reconstruct $\rho = \arg\max_\rho \mathbf{Pr}[y|\rho, A]$, where y is the empirical result of the distribution p. A general prescription for tomography is defined as follows.

1. Prepare target state ρ.

2. Measure the state ρ with different projectors.

3. Obtain a probability distribution of measurement outcomes p, by sampling repeatedly.

4. Use estimator p to reconstruct ρ.

   The apparatus for learning ρ is extremely simple:

   The computational complexity and sampling complexity of quantum state tomography are high. Algorithms exist that optimally choose the set of measurement operators, so that the number of samples needed is minimal. Full quantum tomography typically requires resources that scale exponentially with the number of qubits, in spite of improvements from techniques such as compressed sensing and direct fidelity estimation [312–314].

   Tomography, although recovering full state information, is limited in estimating noise of a process. This is because the state preparation and measurement process can be erroneous as well. This is commonly referred to as the SPAM (State Preparation And Measurement) error. With current technologies, SPAM errors are dominant sources of noise.

8.1.3   RANDOMIZED BENCHMARKING

In contrast, randomized benchmarking (RB) [315–319] emphasizes on gate errors more than SPAM errors by applying a long sequence of gates and postponing measurements to the very end. It can be summarized in the following steps.

1. Prepare target state ρ in computational basis.

2. Select randomly m Clifford gates: $\mathcal{C}$.

3. Apply gates in $\mathcal{C}$ in sequence, and then add the inverse gate at the end.

4. Measure in computational basis E.

5. Obtain estimate $p = \mathbf{Pr}[E|\mathcal{C}, \rho]$.

$$\rho - \boxed{C_0} - \boxed{C_1} - \boxed{C_2} - \cdots - \boxed{C_m} - \boxed{C_{inv}} - \measuredangle$$

Note that we can always find a gate C_{inv} that is the inverse of the product of all Clifford gates being applied. The interpretation of randomized benchmarking is that it estimates the average error rate of average Clifford gates. Notice that RB focuses on gate-independent noise caused not by the Clifford gates we apply, but rather other noise that happens during the process. As we increase the length m, the system will drift farther and farther. This phenomenon is commonly referred to as the "RB decay."

There has been no standard criteria on choosing the circuit length m, the number of random Clifford gate sets, or the number of measurement shots per circuit. Some tools [320] have automated the process of choosing these parameters to ensure getting the most information out of a small number of experiments.

Interleaved Randomized Benchmarking
In order to characterize the error rates of a *particular* gate G, we can choose to interleave G throughout the standard randomized benchmarking sequence [321]. Now the circuit becomes:

$$\rho - \boxed{C_0} - \boxed{G} - \boxed{C_1} - \boxed{G} - \boxed{C_2} - \boxed{G} - \cdots - \boxed{G} - \boxed{C_m} - \boxed{G} - \boxed{C'_{inv}} - \measuredangle$$

Note that we still need to inverse the Clifford sequence. When G is a Clifford gate, we can always find such inverse gate.

Now if we compare the RB decay of this interleaved RB sequence with that of the original standard RB sequence, we expect to obtain the effect of noise from the extra application of G, and thus bound the fidelity of G.

A number of variants of randomized benchmarking have been proposed [322–326]. Each trades, to different degrees, complexity for more comprehensive noise characterizations. Some recent examples include extended randomized benchmarking (XRB) [323, 327] and cycle benchmarking (CRB) [328], which capture the behavior of a quantum system more realistically than the traditional RB does.

8.2   NOISE MITIGATION STRATEGIES

Noise mitigation is one of the biggest challenges facing the QC community. Without strategies to reduce or get around the physical noises, any execution of a quantum program is almost always doomed to fail under such stringent conditions. Noise mitigation techniques work by strategically designing more robust qubits with an ensemble of elements to prolong their lifetimes or performing more accurate gate operations with a composition of pulses to improve their fidelities, etc.

Current NISQ computers [14, 51, 56] lack the ability to isolate and control a sufficiently large number of quantum bits (qubits) with high precision. Scaling up a quantum computer means improvements in both the quantity and the quality of the qubits. On one hand, we want to equip a quantum computer with as many qubits as possible to accommodate large quantum applications. On the other hand, we want each qubit to be as long-lived and controllable as possible to run quantum programs fast and reliably.

8.2.1   RANDOMIZED COMPILING

Quantum gates usually have the distinction between "easy" and "hard" for a given architecture—we define the gates that can be implemented with relatively high precision or low resource cost as easy gates, and the others as hard gates. A typically division for a NISQ computer architecture is the single-qubit gates vs. multi-qubit gates. Single-qubit gates are considered easy because they are usually associated with smaller error rates in a NISQ architecture.

Randomized compiling is a strategy where one inserts random gates into a quantum circuit, and averages over many of those independently sampled random circuits. Remarkably, all coherent errors and non-Markovian noises can thus be converted into stochastic Pauli errors which are arguably easier to detect and correct, while preserving the logical operations of a quantum circuit. While the effect of noise on the individual random circuit may be different, the expected noise on multiple random circuits is scrambled and tailored into a simple stochastic form. The proof of the tailored noise is beyond the scope of the book, but we shall see in this section, how a quantum circuit is compiled into a random one that is more robust to noise.

Specifically, the algorithm goes as follows.

1. Arrange a quantum circuit U into alternating cycles of easy and hard gates: $U = \Pi_{k=1}^{d} U_k G_k$, where d is the number of cycles, G_k (or U_k) is a cycle of easy (or hard) gates on n qubits.

2. Insert a layer of randomly selected Pauli gates on n qubits before the hard gates in each cycle, denoted as $\vec{P}_k = \bigotimes_{i=1}^{n} P_k^{(i)}$, where each $P_k^{(i)}$ is sampled from the Pauli group $P \in \{I, X, Y, Z\}$. In the following, we will omit the implicit superscript and denote the random Pauli gates for the kth cycle as P_k.

3. Insert a layer of correction gates after the hard gates in each cycle, denoted as $\vec{P}_k^c$, such that logical equivalence is preserved for the kth cycle: $U_k = \vec{P}_k^c U_k \vec{P}_k$.

4. Absorb the inserted gates around the easy gates in each cycle into the dressed form $G_k' = \vec{P}_k G_k \vec{P}_{k-1}^c$.

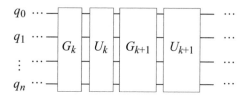

Figure 8.1: An example quantum circuit sliced into interleaved layers of single-qubit gates and two-qubit gates. Shown here the kth cycle and the $(k+1)$th cycle.

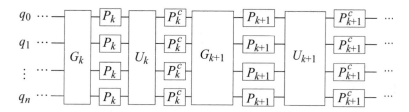

Figure 8.2: Randomly sample Pauli gates on n qubits are inserted before the hard gates, and their corresponding correction gates inserted after the hard gates.

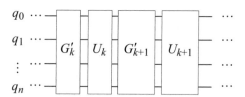

Figure 8.3: The inserted gates are absorbed into the easy gates. As a result, the final random circuit has the same depth as that of the original circuit.

More concretely, suppose the kth cycle consists of a layer of single-qubit gates (as easy gates) denoted as G_k and a layer of two-qubit gates (as hard gates) denoted as U_k, as shown in Figure 8.1. After inserting the random twirling gates and their corrections, we arrive at the circuit in Figure 8.2. Then the random gates are compiled with the easy gates together, as shown in Figure 8.3.

[329] demonstrated that the randomized transformation can be viewed as a scrambling of noise on the quantum circuit. Thus, if we average over multiple shots of randomly sampled circuits, the randomization tailors the noise into stochastic Pauli errors. Consequently, we have obtained a logically equivalent circuit, whose circuit depth is unchanged asymptotically, but is more resilient to external noises. For detailed analysis, we refer the interested reader to the original work in [329].

8.2.2   NOISE-AWARE MAPPING

As much as we want the qubits to be reliable and long-lived, there is still noises present in the qubits themselves and the gate operations we apply on them. With current superconducting technology, it is hard to make identical transmon qubits in a device. Gate errors for different qubits can vary by the hour.

Several recent quantum circuit mapping and scheduling techniques use heuristics to optimize for specific program input, physical machine size and physical topology. Two studies from Princeton [228] and GATech [229] go even further, observing from IBM daily calibration data that qubits and links between qubits vary substantially in their error rate. These compilers take these daily variations into account and optimize to increase the probability of correct program output. A follow-on study [330] demonstrates that this approach substantially reduces error rate as compared to native compilers for the IBM, Rigetti, and UMD quantum machines.

8.2.3   CROSSTALK-AWARE SCHEDULING

In this section, we highlight a number of strategies for reducing crosstalk error in superconducting architectures. A number of hardware features have been proposed to help mitigate crosstalk: (i) connectivity reduction, (ii) qubit frequency tuning, and (iii) coupler tuning. In addition to these hardware features, some software constraints are usually imposed to effectively reduce crosstalk; for example, certain operations may be prohibited to occur simultaneously.

Connectivity reduction works by building devices with sparse connections between qubits, hence reducing the number of possible crosstalk channels. This greatly increases the circuit mapping and re-mapping overhead for executing a logical circuit, since many SWAP gates are needed. Moreover, this model necessitates an intelligent scheduler to serialize operations to avoid crosstalk [232]. This strategy is commonly deployed for fixed-frequency transmon architectures, e.g., from IBM [51]. Because of their non-tunable nature, these architectures have stringent constraints on the initial qubit frequency; a number of optimizers are proposed for this issue [331, 332].

A second class of techniques rely on actively *tuning qubit frequencies* to avoid crosstalk, featured in some prototypes [333] and by Google [334]. Software can decide when to schedule an instruction and which frequency to operate the instruction at. In this class, [335] found a frequency assignment for the surface code circuit; [336] suggests a sudoku-style pattern of frequency assignment for cavity grid.

A third class builds not only frequency-tunable qubits but also *tunable couplers* between qubits, termed "gmon" architectures [337]. Without resorting to permanently reducing device connectivity in hardware, a different subset of connections are activated (via flux drives to the couplers) at different time steps. As such, a schedule for when to activate couplers is needed. For instance, Google proposes a tiling pattern in [66].

Most previous studies on quantum program compilation [159, 305] have largely targeted short program execution time (i.e., low circuit depth), and neglected the impact of gate errors such as crosstalk. Optimizations are performed at the gate level, typically involving strategic qubit mapping and instruction scheduling. Recent efforts [232, 332] are among the first to explore the compiler's role, such as designing intelligent scheduler, to avoid crosstalk.

8.3   QUANTUM ERROR CORRECTION

Quantum error correction (QEC), first developed in [338, 339], achieves fault tolerance by repeatedly discretizing continuous errors into digital errors and using many redundant qubits to flag any errors that have occurred to the quantum state, an idea that is not so distant from memory refresh and classical coding theory. It allows us to track and correct errors in real-time while executing a quantum program. QEC is an extraordinary discovery—it not only explains why we can detect and correct a quantum error, but also provides a recipe for doing so systematically. It is a blueprint for how to build a future large-scale quantum computer fault-tolerantly. Since the focus of the book is on near-term NISQ research, we have been postponing discussions on QEC until now. A single section in the book does not justify how remarkable QEC is; nonetheless, we will demonstrate a selection of fundamental concepts in QEC, first via an example then via a generalized principle. For details about quantum error correction, we refer the readers to [22, 25, 86, 157, 340, 341].

8.3.1   BASIC PRINCIPLES OF QEC

As previously discussed, quantum systems are not ideal. There are many variables that have an impact on the outcome of a computation. The fidelity rates and coherence times are some factors posing challenges. Quantum gate operations and control signals are not perfect. And all of these errors build up to non-negligable amounts. That is why we need a way to correct accumulating errors. This is the motivation behind quantum error correction. Simply put, the purpose of quantum error correction can be summarized as protecting quantum circuits from noise.

Quantum Error vs. Classical Error
Classically, we are using bits, so the information is stored in 0's and 1's. Whenever there is an error, the bit is the opposite of what it is supposed to be (i.e., a bit flip). Because classical errors are just accidental bit flips, they are digitized. However, quantum errors are continuous. This continuous error can be mathematically modeled as follows:

$$|0\rangle \xrightarrow{\text{X gate}} \sqrt{\epsilon}\,|0\rangle + \sqrt{1-\epsilon}\,|1\rangle. \qquad (8.1)$$

Even though physicists do their best to reduce this effect, it sometimes is not enough. There are a lot of questions that rise from this situation. Are we able to detect and measure how big/small this error ϵ is, or even if we can, is it better to correct it right now or later? One of the hard to questions to answer is at what point do we decide to attempt to correct ϵ?

Key Ingredients in Quantum Error Correction

There are two main ideas that make quantum error correction possible. One idea is to use redundant encoding of information, just like in QR codes. This way, effects of noise in certain parts of the system can be tolerated and will not end up corrupting the state of the system. Another main idea is to digitize quantum errors, since we know how to deal with digitized errors, as they resemble the classical case:

- redundancy to encode information and

- digitizing quantum error.

Quantum Error Correction Code (QECC)

Quantum error correction code is a mapping from k logical qubits to n physical qubits. Here, we must emphasize that n is strictly greater than k, as it takes many physical qubits to realize one logical qubit. The idea is to use n physical qubits to encode (protect) k qubits of information. Exactly $n - k$ qubits are used for redundancy. This mapping can be shown as follows:

$$|0_L\rangle = |000\rangle \tag{8.2}$$

$$|1_L\rangle = |111\rangle . \tag{8.3}$$

In the above example, $|0_L\rangle$ stands for the "logical" $|0\rangle$ state of the qubit, and it is realized by three physical qubits. The string 000 and 111 are called the logical *codewords* of the code. Now suppose that with some small probability p, one of the physical qubits flipped, and we got $|001\rangle$. The original "logical" qubit can still be recovered, for example through a majority vote of qubits. We would conclude that the third qubit flipped, and the actual qubit was $|0_L\rangle$.

How to Locate a Bit Flip?

Continuing the above example and representation, locating bit flips can be accomplished by looking at output sequences of a two-qubit operator. These operators are *ZZI* and *IZZ*, and each of the operators act on only one qubit in order. For example, *ZZI* means a Z gate is applied to both the first and the second qubit and the third qubit is left untouched. Recall that

$$Z |0\rangle = |0\rangle \tag{8.4}$$

$$Z |1\rangle = - |1\rangle . \tag{8.5}$$

Now, for a state $|\psi\rangle$ we can look at what the eigenvalues of these two-qubit operators are. And if we apply both of these two-qubit gates consecutively, we can determine which bit flipped. Now suppose that $|\psi\rangle = |100\rangle$. This means,

$$ZZI |100\rangle = - |100\rangle \tag{8.6}$$

$$IZZ \, |100\rangle = |100\rangle \, . \tag{8.7}$$

The eigenvalues observed (in order) are $(-1, +1)$. This sequence tells us that it's the first qubit that is flipped. For instance, if the second qubit was flipped, we would instead observe a sequence that is $(-1, -1)$. Similarly, we would see $(+1, -1)$ if the third qubit was flipped.

If one wishes to compute the phase flip of a qubit, then all Z gates should be replaced by X gates, and all $|0\rangle$ and $|1\rangle$ should be replaced by $|+\rangle$ and $|-\rangle$. This preserves the stabilizer formalism, as the X gate gives $(+1, -1)$ as eigenvalues when it acts on $(|+\rangle, |-\rangle)$. Everything else, just remains the same.

Check Matrix Formalism

In the literature, the extension of how to locate bit flips to a more generalized case comes through the check matrix formalism. The idea of a check matrix is to create a set of qubit operations using the stabilizer formalism, with enough permutations sequences of eigenvalues to determine which qubit is flipped. Each row in the check matrix is a gate operation that needs to be applied to the system, and each column is representative of physical qubits. For example, the check matrix formalism for the above example would contain two rows, one for IZZ, and one for ZZI. It would also contain three columns, as there are three physical qubits in that system. The check matrix would be:

$$\begin{pmatrix} 1 & 1 & 0 \\ 0 & 1 & 1 \end{pmatrix}, \tag{8.8}$$

where 1's stand for Z (for bit flip) or X (for phase flip) gates, and 0's for the identity matrix. This matrix shows that first, ZZI must applied, followed by IZZ. A more complicated example where eight physical qubits are used would be

$$\begin{pmatrix} 1 & 1 & 1 & 1 & 0 & 0 & 0 & 0 \\ 1 & 1 & 0 & 0 & 1 & 1 & 0 & 0 \\ 1 & 0 & 1 & 0 & 1 & 0 & 1 & 0 \end{pmatrix}, \tag{8.9}$$

where we see that a series of three gate operations is necessary to encode enough sequences so that one can distinguish which qubit is flipped.

Nine-qubit Shor Code

Of course, using physical qubits to protect against bit flips would be no use, if one doesn't also protect against phase flips, and vice versa. Unfortunately, each of the physical qubits individually need to also be protected by a second layer of concatenated physical qubits in this case. This gives rise to what is called the nine-qubit Shor Code, a two-layer, 3x3 physical qubit set that protects against both phase and bit flips and encodes one logical qubit. This way, one layer protects against phase flips and the other against bit flips. The logical qubit encoded this way gives us:

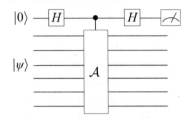

Figure 8.4: The circuit describing a projective measurement for operation A on state $|\psi\rangle$. Here, A can mean any stabilizer n-qubit gate. For example, it can mean *IZZ* described above. There needs to be an additional ancilla qubit for this process.

$$|0_L\rangle = \left[\frac{1}{\sqrt{2}}|000\rangle + \frac{1}{\sqrt{2}}|111\rangle\right]^{\otimes 3} \tag{8.10}$$

$$|1_L\rangle = \left[\frac{1}{\sqrt{2}}|000\rangle - \frac{1}{\sqrt{2}}|111\rangle\right]^{\otimes 3}. \tag{8.11}$$

In this case, operators to check whether phase flips or bit flips occurred changes. We need three sets of bit flip checks, and two sets of phase flip checks. These gates are given below:

$$Z_1 Z_2, Z_2 Z_3 \tag{8.12}$$

$$Z_4 Z_5, Z_5 Z_6 \tag{8.13}$$

$$Z_7 Z_8, Z_8 Z_9 \tag{8.14}$$

$$X_1 X_2 X_3, X_4 X_5 X_6 \tag{8.15}$$

$$X_4 X_5 X_6, X_7 X_8 X_9. \tag{8.16}$$

Projective Measurements

We now describe how to *implement the stabilizer operators* in a quantum circuit. The stabilizer operators (including the example of *ZZI* in the previous section) are implemented as projective measurements. To see why, we consider the circuit below by calculating the quantum state at each time step in the circuit in Figure 8.4.

1. $|0\rangle |\psi\rangle$

2. $|+\rangle |\psi\rangle$

3. $\frac{1}{\sqrt{2}}(|0\rangle |\psi\rangle + |1\rangle A |\psi\rangle)$

4. $\frac{1}{2}[(|0\rangle + |1\rangle) |\psi\rangle + (|0\rangle - |1\rangle) A |\psi\rangle] = |0\rangle \frac{I+A}{2} |\psi\rangle + |1\rangle \frac{I-A}{2} |\psi\rangle$

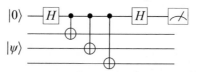

Figure 8.5: Syndrome measurement for the stabilizer operator $X_1 X_2 X_3$.

where operators $\frac{I+A}{2}$ and $\frac{I-A}{2}$ are called "projectors." It can be shown that any arbitrary state $|\psi\rangle$ can be decomposed into orthogonal states as follows:

$$|\psi\rangle = \alpha |\psi_+\rangle + \beta |\psi_-\rangle, \tag{8.17}$$

where $|\psi_+\rangle$ and $|\psi_-\rangle$ are the eigenstates of A with eigenvalues $(+1, -1)$, respectively. In other words,

$$A |\psi_+\rangle = |\psi_+\rangle, A |\psi_-\rangle = -|\psi_-\rangle. \tag{8.18}$$

One can think of these states as "no error" and "error" states as well. Since when we have no error, stabilizer operators give us an eigenvalue of $+1$, and when we have error, it's -1. Therefore, we can see further that

$$\frac{I + A}{2}(\alpha |\psi_+\rangle + \beta |\psi_-\rangle) = \alpha \frac{1+1}{2} |\psi_+\rangle + \beta \frac{1-1}{2} |\psi_-\rangle = \alpha |\psi_+\rangle, \tag{8.19}$$

which shows us that we recover the original "no error" state $|\psi_+\rangle$ with probability α and the "error" state $|\psi_-\rangle$ with probability β. So if $\alpha \sqrt{1-\epsilon}$ and $\beta \sqrt{\epsilon}$ where $\epsilon \ll 1$ (i.e., the error is small), we recover the "no error" state with high probability. This procedure shows that "projectors" actually transform the arbitrary state $|\psi\rangle$ into one of the two states, the "no error" and the "error" states.

For instance, measuring the stabilizer operator $X_1 X_2 X_3$ means that we perform a projective measurement using an ancilla qubit, as shown in Figure 8.5.

8.3.2   STABILIZER CODES

Using the stabilizer formalism defined earlier, we can construct a family of quantum error correction codes, defined by. Thanks to the simplicity in the formalism, we are able to borrow the concepts from linear codes in classical coding theory.

Definition 8.6   An $[n, k]$ *stabilizer code* $C(S)$ is defined as the vector space stabilized by the operators from the abelian subgroup $S = \langle g_1, g_2, \ldots, g_{n-k} \rangle$, where $g_i \in P_n \setminus -I$ is a stabilizer from the n-qubit Pauli group, represented as a length-n Pauli string. k is the number of logical qubits that $C(S)$ encodes.

$$C(S) = \{|\psi\rangle \in \mathcal{H}, \text{s.t. } g |\psi\rangle = |\psi\rangle \, \forall g \in S\}.$$

The nine-qubit Shor code can therefore be written as a [9, 1] stabilizer code, which uses 9 physical qubits to encode one logical qubit where the stabilizers are Equations (8.12)–(8.16).

With this definition of a quantum error correction code, we have the following theorem showing the set of errors that can be corrected by the stabilizer code.

Theorem 8.7   *Given a set of Pauli errors $\mathcal{E}$, if for all $E_i, E_j \in \mathcal{E}$, $\exists g \in S$, s.t. $E_i^\dagger E_j g = -g E_i^\dagger E_j$, then the set of errors $\mathcal{E}$ is correctable by the stabilizer code $C(S)$.*

The proof of this theorem can be found in [86]. Effectively, if a Pauli error anti-commute with a stabilizer, then the stabilizer can detect and correct an occurrence of the error. This is because, upon projective measurement of g, we can observe that $E_i |\psi\rangle$ is projected to the -1 eigenstate of g, indicating the occurrence of error. The series of projective measurement outcomes are called the *syndrome*. As such, each error will leave a signature in the syndrome. To tell two errors apart, we need their syndromes to be distinct. This process is called *decoding* of the syndromes. We can thus correct the errors appropriately.

8.3.3   TRANSVERSALITY AND EASTIN–KNILL THEOREM

Once the error correction code is defined, the next step is to define how to implement logical operations on the codewords of the code fault-tolerantly. After all, we need to prevent errors from propagating through computation. For each error correction code, there is a class of gates whose logical gate operations (i.e., encoded gates) are easy to implement fault-tolerantly, namely the *transversal quantum gates*. To prevent the propagation of error during a logical operation, we can impose the requirement that each physical gate for the logical gate acts on at most one physical qubit in each of the n-qubit code block using the $[n, k]$ code. For example, if logical Hadamard gate consists of physical Hadamard gate on each of the n physical qubits: $\tilde{H} = \bigotimes_{i=1}^{n} H$, it would be considered as a transversal Hadamard. In the case of stabilizer codes, for example, the logical X (i.e., $|0\rangle_L \rightarrow |1\rangle_L$) and logical Z (i.e., $|1\rangle_L \rightarrow -|1\rangle_L$) operations can be derived by finding the operator $h \in P_n \setminus S$ but commutes with all $g \in S$. Other logical operations are potentially more difficult to implement.

Transversal gates are preferable because the noisy, physical gates are localized in each code block, preventing errors from spreading uncontrollably through computation. However, the *Eastin–Knill theorem* states that no quantum error correction code can transversally implement a universal gate set. So we have to circumvent the theorem using other techniques to implement fault-tolerant quantum gates. We motivate a class of such techniques by the Knill's error correction picture using gate teleportation.

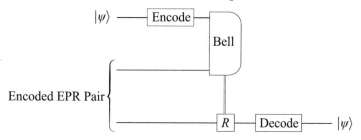

Figure 8.6: An encoded teleportation circuit. The input to the teleportation circuits is the encoded qubits and the encoded EPR pair; the circuit consists of the encoded Bell-basis measurement and encoded recovery Pauli operators.

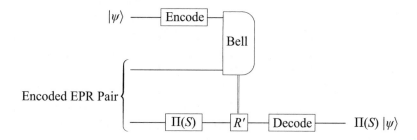

Figure 8.7: Circuit diagram for teleporting a projective measurement $\Pi(S)$. The output of the circuit is $R\Pi(S)|\tilde{\psi}\rangle = \Pi(S)R'|\tilde{\psi}\rangle$.

8.3.4   KNILL'S ERROR CORRECTION PICTURE

The Knill's error correction picture differs from conventional error correction in many ways; we highlight one of the differences, namely the concept of error correcting teleportation [342], which generalizes from gate teleportation [234]. In this picture, error correction is combined with logical gate into one step, instead of the conventional syndrome-based scheme discussed earlier. The error correcting teleportation circuit, uses a generalization of the teleportation circuit to use encoded states and encoded gates, as shown in Figure 8.6.

The key observation in Knill's error correction picture is that a stabilizer projection on the encoded qubits before teleportation is equivalent to a stabilizer projection after the teleportation (up to Pauli modification due to the recovery operator at the end of the teleportation circuit), as shown in Figure 8.7.

The argument follows similarly from that of the gate teleportation technique for unitary gates, but generalized to projective measurements. Consequently, the syndrome of the input state can be deduced from the teleportation Bell measurement [342].

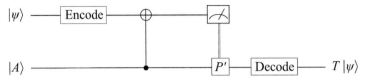

Figure 8.8: The role of magic states in the Knill error correction picture. The magic state $|A\rangle$ is teleported to implement a T gate fault-tolerantly. Here, $P' = TXT^\dagger = XS^\dagger$. All gates (i.e., CNOT gate, P' gate, and measurement) are implemented fault-tolerantly.

Magic State Distillation

We can use this picture of error correction to motivate a technique that aims to remedy the Eastin–Knill theorem, namely *magic state distillation*. One of the advantages of the error correcting teleportation picture [233, 342] is that the difficulty in performing a logical gate U fault-tolerantly on the encoded $|\psi\rangle$ is shifted to the difficulty in preparing an encoded resource state (also known as magic states) fault-tolerantly. The latter is in general easier, because the preparation of specific magic states is generally easier than performing an operation on an unknown state: the magic state can be prepared offline (i.e., prior to the computation), and we can discard a resource state and start over in case that the preparation circuit fails.

Magic state distillation, first proposed by Bravyi and Kitaev [28], is precisely the process for preparing a resource state fault-tolerantly that corresponds to a non-transversal gate for the error correction code.

A widely studied magic state is the resource state for T gate (i.e., $\pi/8$ gate). For example, we can input the following resource state:

$$|A\rangle = T\,|+\rangle = \frac{1}{\sqrt{2}}\left(e^{-i\pi/8}\,|0\rangle + e^{i\pi/8}\,|1\rangle\right)$$

to the (single-qubit version) teleportation circuit, as shown in Figure 8.8.

As a result, the problem of implementing the non-transversal T gate fault-tolerantly is reduced to the problem of preparing (in advance) high-fidelity magic states.

T Gates in Quantum Algorithms

To put the cost of QEC in perspective, we examine the overhead of non-transversal operations such as S and T gates. S and T gates are important operations in many useful quantum algorithms, and their error-corrected execution requires magic state resources. When the number of T gates in an application is low, the circuit is in fact able to be efficiently simulated classically [180]. T gates have been shown to comprise between 25% and 30% of the instruction stream of useful quantum applications [343]. Others claim even higher percentages for specific application sets, of between 40% and 47% [344].

For an estimate of the total number of required T gates in these applications, take as an example the algorithm to estimate the molecular ground state energy of the molecule Fe_2S_2. It requires approximately 10^4 iteration steps for "sufficient" accuracy, each comprised of 7.4×10^6 rotations [345]. Each of these controlled rotations can be decomposed to sufficient accuracy using approximately 50 T gates per rotation [215]. All of this combines to yield a total number of T gates of order 10^{12}. As a result, it is crucial to optimize for the resource overhead required by the execution of T gates at this scale to ensure the successful execution of many important quantum algorithms.

Bravyi–Haah Distillation Protocol

The last piece in the problem is how to prepare the resource states fault-tolerantly. Distillation protocols are circuits that accept as input a number of potentially faulty raw magic states, use some ancillary qubits, and output a smaller number of higher fidelity magic states. The input-output ratio, denoted as $n \to k$, assesses the efficiency of a protocol. Below we focus on a popular, low-overhead distillation protocol known as the Bravyi–Haah distillation protocol [346].

To produce k magic states, Bravyi–Haah state distillation circuits take as input $3k + 8$ low-fidelity states, use $k + 5$ ancillary qubits, and k additional qubits for higher-fidelity output magic states, thus denoted as the $3k + 8 \to k$ protocol. The total number of qubits involved in each of such circuit is then $5k + 13$, which defines the area cost of the circuit module.

The intuition behind the protocol is to "make good magic states from many bad ones." Given a number of low-fidelity states, the protocol uses a syndrome measurement technique to verify quality, and discards states that are bad. Then, the circuit will convert the subset of good states into a single qubit state. The output magic states will have a suppression of error, only if the filtering and conversion follows a particular pattern. This is specified by the parity-check matrix in the protocol. Notably, if the input (injected) states are characterized by error rate ϵ_{inject}, the output state fidelity is improved with this procedure to $(1 + 3k)\epsilon_{inject}^2$. Due to the filtering step, the success probability of the protocol is, to first order, given by $1 - (8 + 3k)\epsilon_{inject} + \cdots$.

8.4 SUMMARY AND OUTLOOK

Quantum computation requires the mitigation of errors caused by imprecise controls and natural decoherence. Noise mitigation and error correction techniques are critical in making quantum computation practical, especially in the NISQ era. In this chapter, we introduced the leading techniques for characterizing and modeling noises, leading to better understanding of the effects of quantum noise. We also described several physical-level noise mitigation techniques for reducing crosstalk and coherent errors. And finally, we illustrate the basics of quantum error correction codes which have the potential to fault-tolerantly execute arbitrary quantum computation.

Further Reading

At the physical level, two widely used noise mitigation techniques include composite pulses [298] (for systematic errors) and dynamical decoupling [347] (for coherent dephasing errors). Scalability remains a challenge in noise mitigation methods. A quantum system needs more intelligent methods for calibration, as it is infeasible to exhaustively calibrate each and every qubits as every coupling of two qubits. It remains an open problem to characterize realistic quantum noises [328, 348], and classically simulate them [349–351].

Quantum error correction is a beautiful mathematical proposal for scaling up quantum computation with fault-tolerance. The capstone of QEC is the theorem called "threshold theorem" [352]. It is one of the most remarkable results of QEC which states, at a high level, that if the physical error rate p is less than some threshold p_{th}, then we can perform fault-tolerant quantum computation to accuracy ϵ with only a moderate increase in circuit size (that is poly-logarithmic in the accuracy ϵ) by concatenation of QEC, and given reasonable assumptions about the physical noise model. The details of the theorem is out of the scope of this review. We refer readers interested in QEC to [23, 25, 86]. In the near term, numerous efforts have been put in designing low-overhead quantum error correction codes adapted to different noise models and program characteristics, as well as aiming to reduce the cost of magic state distillation protocols.

CHAPTER 9

Classical Simulation of Quantum Computation

Last but not least, we discuss the classical simulation of quantum computation, where we explore the techniques for efficiently simulating quantum circuits on a classical computer. This is an important subject, in part because it allows us to execute a quantum program and verify its correctness even when no quantum hardware is available. It is thus an essential tool for testing and debugging quantum programs. But it also sheds light on the not-so-well-understood computational power boundary between classical computers and quantum computers. We would like to understand how much of quantum processes (if not all) can be efficiently simulated on a classical computer. In other words, understanding various classical simulation techniques can give us insights on what are the key ingredients in quantum computing that bring the advantage in computing power. For instance, is entanglement responsible for quantum speedup, or is there more to it? Those are the kind of questions raised and hopefully answered when we study the simulation of quantum computation. After defining what classical simulation means, we direct the reader's attention to some leading techniques, namely simulation using density matrices, stabilizer formalism, and graphical models.

9.1 STRONG VS. WEAK SIMULATION: AN OVERVIEW

The aim of a classical simulation is to mimic the dynamics of a quantum systems, and accurately reproduce the outcomes of a quantum circuit. Recall from earlier chapters, for every execution of a quantum circuit on a quantum computer, we obtain a sample bit-string from a probability distribution resulting from the measurements at the end of its circuit. More specifically, given a quantum circuit U, with n input qubits and N number of gates. For an efficient quantum circuit, N is usually $O(poly(n))$. Upon measuring the qubits, we read out a sample bit-string, denoted as $\alpha \in \{0, 1\}^n$, with probability:

$$P(\alpha) = |\langle \alpha | U | 000\dots0 \rangle|^2.$$

Now, a quantum computer can report the result from measuring qubits at the end; we are interested in doing it classically. In general, a classical computer have two options, namely to *strongly* or *weakly* simulate the resulting probability distribution.

Definition 9.1   *Strong simulation* aims to *calculate the probabilities* of the output measurement outcomes efficiently with high accuracy using a classical computer.

With strong simulation, a classical computer needs to show that it has reproduced the outcome of a quantum circuit by writing down the probabilities of some or all of the possible bit-strings. To verify, we can run the quantum computer multiple times and check if its sampling probability is close to the reported value by the classical computer. More specifically, we emphasize that there are typically two styles of strong simulations:

1. "Evaluating all amplitudes"—we aim to calculate $P(\alpha)$, $\forall \alpha$ and

2. "Evaluating one amplitude"—we calculate the probability of one of the outcomes, e.g., $P(0 \ldots 0)$.

With weak simulation, a classical computer performs as a sampling device, acting more closely as a quantum computer does. In particular, we ahve the following Definition.

Definition 9.2   *Weak simulation* aims to *sample once* from the output distribution efficiently using a classical computer.

Each time you simulate a circuit weakly, you will obtain an outcome α with probability according to a distribution close to the true probability distribution $P(\alpha)$ had one done it quantumly.

We remark that strong and weak simulations are fundamentally *different* notions. In other words, we can find some quantum circuits that are trivially simulable weakly, but are unlikely to be efficiently simulable strongly. For example, [353] shows that strong simulation of some quantum circuit is #P-complete.

Furthermore, *strong simulation implies weak simulation.* The forward direction is simple: if the probabilities are calculated, then you can sample according to the probabilities. But if you can sample once in poly time and there are exponential possibilities, it is not immediately clear how to recover all amplitudes with accuracy. Techniques developed for classical simulation have been focusing on simulating quantum circuit strongly. However, weak simulation is closer to what we are interested in physically, because a quantum device produces a sample at a time upon measurements. Strong simulation, especially for evaluating all amplitudes, may after all be too harsh on the classical computers.

Despite the common belief that classical simulation of universal quantum circuits does not scale well, efficient simulations for some restricted classes of circuits exists. One such example in this book is the Clifford circuits, which have been shown to be efficiently simulable using the stabilizer formalism. These restricted classes also tell us what features in quantum circuits enable the quantum computational power.

9.1.1   DISTANCE MEASURES

To evaluate how well a classical simulation reproduced the probability distribution of a quantum circuit, we define some relevant distance measures for classical probability distributions. There are many different distances used for comparing two probability distributions p and q; we mostly follow the analysis in [354, 355]. Consider two discrete probability distributions $p = (p_1, \ldots, p_d)$ and $q = (q_1, \ldots, q_d)$ over the same space Ω where $|\Omega| = d$.

Definition 9.3   The *total variation distance* between p and q is defined as

$$d_{TV}(p, q) = \frac{1}{2} \sum_{i=1}^{d} |p_i - q_i| = \frac{1}{2} ||p - q||_1.$$

The total variation distance, which takes value between 0 and 1, measures the worst probability discrepancy between a sample from p and a sample from q, i.e., $d_{TV}(p, q) = \max_{x \in \Omega} |\mathbf{Pr}_p[x] - \mathbf{Pr}_q[x]|$.

Definition 9.4   The ℓ_2 *distance* between p and q is defined as

$$d_{\ell_2}(p, q) = \left(\sum_{i=1}^{d} (p_i - q_i)^2 \right)^{1/2} = ||p - q||_2.$$

The ℓ_2 distance, which takes value between 0 and $\sqrt{2}$, is related to the total variation distance by $d_{\ell_2}(p, q) \leq 2d_{TV}(p, q) \leq \sqrt{d} d_{\ell_2}(p, q)$.

Definition 9.5   The *Hellinger distance* between p and q is defined as

$$d_H(p, q) = \left(\sum_{i=1}^{d} (\sqrt{p_i} - \sqrt{q_i})^2 \right)^{1/2}.$$

The Hellinger distance, which take value between 0 and $\sqrt{2}$, is related to the total variation distance by $d_H^2(p, q) \leq 2d_{TV}(p, q) \leq 2d_H(p, q)$.

For completeness, we should note that many of the distance metrics above have a quantum analogue. The goal of distance measures for quantum mixed states is to quantify by how much do the quantum probability distributions of the two quantum states differ—see Chapter 2 for a short review on quantum probability.

Definition 9.6   The *trace distance* between two mixed states ρ and σ is defined as

$$D_{tr}(\rho, \sigma) = \frac{1}{2}||\rho - \sigma||_1 = \frac{1}{2}tr\left(\sqrt{(\rho - \sigma)^\dagger(\rho - \sigma)}\right).$$

The trace distance, which takes value between 0 and 1, can be viewed as the quantum analogue of the total variation distance, in that D_{tr} calculates the maximum probability that two states ρ and σ can be discriminated by measurements.

Definition 9.7   The *Hilbert–Schmidt distance* between ρ and σ is defined as

$$D_{HS}(\rho, \sigma) = ||\rho - \sigma||_F = tr\left((\rho - \sigma)^2\right)^{1/2},$$

where $||\cdot||_F$ is also called the Frobenius norm. The Hilbert–Schmidt distance is the quantum analogue of the ℓ_2 distance. It relates to the trace distance by $D_{HS}(\rho, \sigma) \leq 2\,D_{tr}(\rho, \sigma) \leq \sqrt{d}\,D_{HS}(\rho, \sigma)$.

Definition 9.8   The *Bures distance* between ρ and σ is defined as

$$D_B(\rho, \sigma) = (2(1 - F(\rho, \sigma)))^{1/2},$$

where $F(\rho, \sigma) = ||\sqrt{\rho}\sqrt{\sigma}||_1$ is the fidelity between the two mixed states ρ and σ. The Bures distance is the quantum analogue of the Hellinger distance. It relates to the trace distance by $D_B^2(\rho, \sigma) \leq 2D_{tr}(\rho, \sigma) \leq 2D_B(\rho, \sigma)$.

With the distance metrics defined, we are going to introduce a number of simulation techniques. In particular, they are all strong simulation techniques. In fact, most simulations that has been developed are strong simulations, and it remains an exciting open problem to explore the possibility in the weak simulation of quantum circuits.

The leading simulation techniques we choose to cover include: density matrix simulation, stabilizer formalism, tensor networks, and undirected graphical model.

9.2   DENSITY MATRICES: THE SCHRÖDINGER PICTURE

In the Schrödinger picture, evolution of quantum systems can be described by tracking its *quantum state* over time, denoted as a time-dependent ket vector $|\psi(t)\rangle$ (or density matrix $\rho(t)$).

Through time, a quantum state is evolved to another by some unitary transformation U:

$$|\psi(0)\rangle \xrightarrow{time} |\psi(t)\rangle = U |\psi(0)\rangle.$$

As such, one of the most straightforward way for simulating quantum circuits is explicitly tracking the transformation of the qubits (in the form of state vector or density matrix) over time. More specifically, we can just compute the state vector or density matrix at each stage of the circuit, by multiplying it with the unitary matrix of the gate one at a time. As shown in Chapter 2, density matrix simulation is needed for quantum circuits with intermediate measurements (which transforms pure quantum states to mixed quantum states), because tracking the state vector over time is not enough for representing mixed state. Furthermore, noisy quantum circuits often need density matrix simulations. In fact, when we analyze the quantum algorithms in Chapter 3, we have been using this strategy to compute the output of the algorithm, acting as a strong simulator by tracking the quantum state over time. From this point forward in this section, for simplicity, we are mostly concerned with using state vectors to simulate a quantum circuit U strongly.[1] For an n-qubit quantum circuit, suppose we start with $|0\ldots0\rangle$ state and want to calculate the probability of an outcome $x \in \{0, 1\}^n$:

$$p(x) = |\langle x|U|0\ldots0\rangle|^2.$$

The circuit U is represented by m quantum gates, each acting on n qubits, so $U = U_m \cdots U_2 U_1$. Each U_i is a $2^n \times 2^n$ matrix with complex entries.

Let us analyze the space and time cost of naively applying the matrix multiplication. To begin with, the size of a state vector (i.e., dimension of a vector) grows exponentially with the number of qubits to 2^n, so there are 2^n number of complex amplitudes to keep track of. For a depth m circuit, a naive strategy needs to apply matrix multiplication to the state vector m times. If the unitary gate is a $2^n \times 2^n$ matrix (i.e., a n-qubit gate), using naive matrix multiplication, we are performing about $O(2^{2n})$ multiplications of two complex amplitudes, and summing them row by row. So the naive algorithm requires a time cost of $O(m2^{2n})$. Every iteration, we need space for 2^{2n} complex numbers (stored with some floating point precision) and $2 \cdot 2^n$ numbers for the input and output state vector. Overall, the space cost is also $O(m2^{2n})$. This is a daunting scaling, both in time and space, for a classical computer. Suppose we can efficiently synthesizing the n-qubit unitary to a sequence of one- and two-qubit gates of length L, one can reduce the space cost to $O(L + 2^n)$ and time cost to $O(L2^n)$, still exponential in the number of qubits. Further optimizations such as data compression, distributed algorithm have been applied to reduce the cost on a classical computer [356, 357]. Due to this exponential scaling in space and time cost, simulations for more than 65 qubits are shown to be challenging, even on today's state-of-the-art supercomputers.

---

[1]In general, density matrices are necessary if intermediate measurements or noise are present in the target quantum circuits.

Simulating Product States

Let us now analyze a special case where the quantum states are product state throughout the circuits. As we will see in the following argument, if we are willing to forgo quantum entanglement, then a classical computer can simulate the quantum circuit efficiently. Suppose we have a product quantum state:

$$|\psi\rangle = |q_1\rangle \otimes |q_2\rangle \otimes \ldots \otimes |q_n\rangle,$$

where each qubit carries two amplitudes, e.g., $|q_1\rangle = \alpha_1 |0\rangle + \beta_1 |1\rangle$. There are only $2n$ number of amplitudes to keep track of, rather than all 2^n of them. As for the unitary matrix, note that being in product state also means that if the gate is local you only need to update the local bit of the amplitudes. (To see why you only need to update locally, note that you are now storing the amplitudes in the above way with $2n$ complex numbers, instead of the original ket vector.) As such, we have seen that entanglement is a key resource for quantum computing power. It is important to note here, though, that it is not sufficient. Note that entanglement is not a sufficient condition, as there exists circuits that entangle all qubits, yet still are efficiently simulable on a classical computer [358–361].

Sum-Over-Path Approach

Using the path integral technique introduced in Chapter 3, we will show that we can simulate a quantum circuit with *polynomial* space on a classical computer, at the cost of (possibly exponential) time complexity. In particular, in the context of computational complexity as defined in Chapter 3, the statement is equivalent to $\text{BQP} \subseteq \text{PSPACE}$ (polynomial space). To begin with, we rewrite the matrix multiplication with the sum-over-paths technique:

$$\langle x|U|0\ldots0\rangle = \sum_{x_i \in \{0,1\}^n, i \in \{1\ldots m\}} \langle x|U_m|x_{m-1}\rangle \langle x_{m-1}|U_{m-1}|x_{m-2}\rangle \cdots \langle x_1|U_1|0\rangle.$$

In total, there are $2^{n(m-1)}$ number of paths to sum over, leading to a time cost of $O(m2^{n(m-1)})$. Due to the tree structure in the sum-over-path construction, we have to store $nm \log m$ number of complex numbers along the paths. It is important to not neglect the space cost for storing the unitaries of the quantum circuit. We again resort to the argument on efficient synthesis of the $2^n \times 2^n$ sized U to a sequence of single- and two-qubit gates, so that $\langle x_i|U_i|x_{i-1}\rangle$ in the summation can be efficiently computed by storing the four complex numbers in U_i, requiring space polynomial in the precision of the complex number. So the space cost overall is reduced to a reasonable $O(nm \log m)$. This concludes our proof of $\text{BQP} \subseteq \text{PSPACE}$.

9.3   STABILIZER FORMALISM: THE HEISENBERG PICTURE

In the Heisenberg picture, evolution of a quantum system can be described by tracking its *operators* over time. Suppose we are consider an operator that is an observable A. In the old

Schrödinger picture, we calculate the expected value of the observable by conjugating the final state vector: $\langle A \rangle = \langle \psi(t)|A|\psi(t) \rangle$. In contrast to the Schrödinger picture in which the state vector changes over time, the Heisenberg picture is an equivalent formalism of quantum mechanics in which the state vector is kept constant at the initial value $|\psi\rangle = |\psi(0)\rangle$, and the operator evolves over time: $\langle A \rangle = \langle \psi(0)|U^{\dagger}(t)AU(t)|\psi(0) \rangle = \langle \psi(0)|A(t)|\psi(0) \rangle$. In other words, we can track the evolution of a quantum system by its time-dependent operator $A(t)$:

$$A(0) \xrightarrow{\text{time}} A(t) = U^{\dagger}A(0)U.$$

Classical simulation based on stabilizer formalism corresponds to the Heisenberg picture of quantum computation. We have seen from Chapter 8 the stabilizer formalism as useful for quantum error correction; now we shall see its application in classical simulation of a subclass of quantum circuit by the *Gottesman–Knill theorem*. Before we start our discussion, we want to emphasize that we are now considering *efficient* simulations of a *restricted class* of quantum circuits. When we say "efficient," we mean $O(poly(n))$ space and time costs, where n is the number of qubits, and "restricted class" here means the stabilizer circuits (also known as the Clifford circuits).

The key idea is to find a compact representation of a quantum state, together with an efficient update rule for transforming the quantum states. We will start by defining the class of circuits in consideration.

Definition 9.9   A quantum gate is a *stabilizer gate* if it is generated from the Clifford group $S = \langle \text{CNOT}, \text{H}, \text{S} \rangle$. In other words, it is a product of $g \in S$.

For example, all Pauli gates belong to this set: $X = HZH$, $Y = iXZ$, $Z = SS$. Notice that a stabilizer gate S conjugates a gate from the Pauli group back to the Pauli group: $SP_iS^{\dagger} = P_j$ up to a phase factor, where $P_i, P_j \in \mathcal{P}$.

Definition 9.10   A state is a *stabilizer state* if it can be prepared from $|00\ldots0\rangle$ using stabilizer gates.

For example, we can list the single-qubit stabilizer states (6 of them): $|0\rangle, |1\rangle, |+\rangle, |-\rangle, |+i\rangle, |-i\rangle$, where $|\pm\rangle = \frac{|0\rangle \pm |1\rangle}{\sqrt{2}}$ and $|\pm i\rangle = \frac{|0\rangle \pm i|1\rangle}{\sqrt{2}}$.

Definition 9.11   A quantum circuit is called a *stabilizer circuit* if it is made of stabilizer gates applied on input state $|00\ldots0\rangle$, and measurements in the computational basis.

Definition 9.12 $|\psi\rangle$ is stabilized by a quantum circuit U, if $U|\psi\rangle = |\psi\rangle$.

We consider a few examples where the states are stabilized by the Pauli gates.

- *I* stabilizes everything. $I|\psi\rangle = |\psi\rangle$.

- X stabilizes $|+\rangle$. $X|+\rangle = X(\frac{|0\rangle+|1\rangle}{\sqrt{2}}) = \frac{|1\rangle+|0\rangle}{\sqrt{2}} = |+\rangle$.

- $-|Z\rangle$ stabilizes $|1\rangle$. $-Z|1\rangle = -(-|1\rangle) = |1\rangle$.

- $X \otimes I = XI$ stabilizes $|+\rangle \otimes |0\rangle$, $|+\rangle \otimes |+\rangle$, $\ldots$.

In order to *uniquely* represent a state, say $|\psi\rangle = |+\rangle \otimes |0\rangle$, we can use its stabilizer(s). In the example $|\psi\rangle$, we have already found it to be stabilized by XI. Since there are multiple states that are stabilized by XI, we ought to find its other stabilizers in order to avoid ambiguity. In other words, we want to find a set S, such that $|\psi\rangle$ is uniquely determined when $s \in S$ simultaneous stabilize $|\psi\rangle$.

In the example $|\psi\rangle = |+\rangle \otimes |0\rangle$, this set is $\{II, XI, IZ, XZ\}$. Note that II stabilizers all two-qubit state, and XZ is the product of XI and IZ. So essentially, to uniquely represent our $|+\rangle \otimes |0\rangle$ state, we only need to keep track of the *two* stabilizers XI and IZ. Remarkably, the Gottesman–Knill theorem says that the number of stabilizers we need to keep track of is only $O(n)$.

Theorem 9.13   Gottesman–Knill theorem *[362] states that there exists classical algorithm that simulates any stabilizer circuit in polynomial time.*

In simulation, we do not need to keep track of the amplitudes of state vector anymore; rather we can keep track of the stabilizer operators. Let us now examine how to update the stabilizer group when applying a quantum gate:

$$|+\rangle \otimes |0\rangle \xrightarrow{I \otimes H} |+\rangle \otimes |+\rangle .$$

We can find the set of operators that simultaneously stabilizes the initial state $|+\rangle |0\rangle$ and the final state $|+\rangle |+\rangle$, respectively:

$$\{II, XI, IZ, XZ\} \xrightarrow{I \otimes H} \{II, XI, IX, XX\}.$$

More compactly, we can write only the generators (i.e., the minimal subset such that every element in the set can be obtained from product of the generators): $\langle XI, IZ \rangle \xrightarrow{I \otimes H} \langle XI, IX \rangle$.

In general, we can track how the quantum systems evolve over time by updating its stabilizer operators:

$$S \xrightarrow{U} USU^\dagger.$$

For convenience, we list the update rules for some common Clifford gates:

- H gate: $X \to Z$, $Z \to X$:

- S gate: $X \to Y$, $Z \to Z$; and

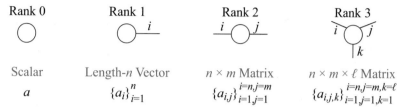

Figure 9.1: Graphical representation of tensors and their mathematical definitions.

- CNOT gate: $XI \rightarrow XX$, $IX \rightarrow IX$, $ZI \rightarrow ZI$, $IZ \rightarrow ZZ$

To complete the explanation of the theorem, we need a few other ingredients [363]: (i) proving that the size of stabilizer generators scales linearly with the number of qubits; (ii) demonstrating that measurement is efficient; and (iii) showing that the amplitude $\langle x|U|00\ldots0\rangle$ can be computed efficiently.

Indeed, one can use the tableau representation to accomplish all of the above-mentioned tasks. The tableau representation is an $(\ell \times 2n)$ matrix storing information about the stabilizer generators, where the number of stabilizer $\ell \in O(n)$. One can also show a $O(poly(n))$ time procedure to update the tableau for measurements in the computational basis. Details of the proof can be found in [363].

9.4 GRAPHICAL MODELS AND TENSOR NETWORK

A different class of simulation techniques is developed based on graphical models. By converting a quantum circuit into a graphical representation, one's hope is to more efficiently perform the transformation $|\psi(0)\rangle \xrightarrow{U} |\psi(t)\rangle$. Let us begin with a technique called *tensor network* simulation.

A rank-k tensor is a k-dimensional matrix. A rank-k tensor is a mathematical object where an entry in the object is located by k indices. Note that in the graphical representation, a tensor is a vertex, and the rank of the tensor is represented as the number of edges connecting to the vertex. In other words, each edge represent an index. We can label the edges with the name of the index. For example, a rank-0 tensor is just a scalar; a rank-1 tensor is a vector (indexed by the position in the vector); a rank-2 tensor is a matrix (indexed by row number and column number); a rank-3 tensor is some data structure that is indexed by 3 numbers, e.g., a movie ticket (indexed by screen number, row number, seat number). In the context of quantum computation, we can make the following equivalence in Figure 9.1.

Now we illustrate how to transform a quantum circuit into a graph of tensors. To map from a quantum circuit to a *tensor network*, we highlight the following correspondence.

- Qubit state: vector $\rightarrow$ 1-d tensor.

- Single-qubit gate: 2×2 matrix (i.e., qubit input index (column) and qubit output index (row)) $\rightarrow$ 2-d tensor.

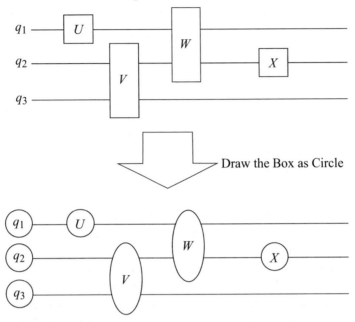

Figure 9.2: Converting from a quantum circuit to a tensor network.

- Two-qubit gate: instead of a 4 × 4 matrix, we can index an entry by 4 indices, namely the qubit 1 input, the qubit 1 output, the qubit 2 input, and the qubit 2 output → 4-d tensor.

Therefore, we can straightforwardly convert a quantum circuit into a network of tensors of various rank, as shown in Figure 9.2.

Furthermore, the final simulation output $\langle \psi | U | 00 \ldots 0 \rangle$ is computed by the product of the final quantum state $\langle \psi |$, the quantum circuit U, and the initial state $| 00 \ldots 0 \rangle$. In terms of the tensor network, the final product is represented by sticking rank-1 tensors at the beginning and end of the network derived from Figure 9.2.

Tensor Network Contraction

Tensor contraction is a process where we merge two tensors into one, absorbing the common edges (i.e., the common indices) between the two tensors. When two tensors share a common index, we can contract the corresponding edge by summing over all possible values of that index. Indeed, tensor network contraction can be thought of as the multi-dimensional generalization of matrix multiplication.

Given two matrices (i.e, rank-2 tensors) A and B, we contract the two tensors using the definition of matrix multiplication. Let us denote A_i^j as ith row, jth col of matrix A; similarly for B. Now, we can contract the edge j:

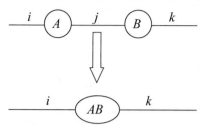

Figure 9.3: Contracting two rank-2 tensors, A and B, is equivalent to the matrix multiplication $C = AB$.

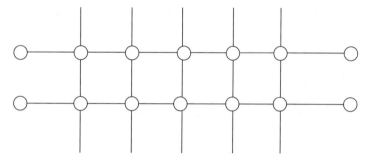

Figure 9.4: Part of a generic tensor network, consisting of ten rank-4 tensors and four rank-1 tensors.

$$\sum_j A_i^j B_j^k = C_i^k.$$

Note that the end result is another matrix C indexed by i and k.

Complexity of Contraction

Contracting one edge take $O(\exp(d))$ time where d is the max rank of tensors involved. To see this, we notice that in our case, each index can take values 0 or 1, so contracting the edge corresponding to that index will yield a summation with two terms. Now consider the case where two tensors are connected by k edges, combining these two tensors would mean summing over d different indices each of which can take two values. So the summation has 2^d terms in total.

Normally, in simulation of quantum circuit, the goal is to contract a tensor network into a single rank-0 tensor. Consider part of a tensor network shown in Figure 9.4. The first problem we encounter is deciding the order of edges to contract. Due to the structure of the network, some ordering may have lower overall cost than that of the other.

Imagine one could contract some tensors in parallel, it is therefore the maximum rank of tensors you encounter during the process of contraction that determines the complexity. So, can

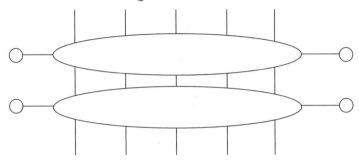

Figure 9.5: First strategy of contraction that results in two rank-12 tensors and four rank-1 tensors. Then contracting the two rank-12 tensors involves contracting 5 edges at once, by summing over 2^5 terms.

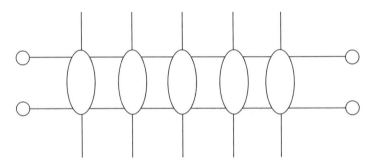

Figure 9.6: Second strategy of contraction that results in five rank-6 tensors and four rank-1 tensors. Then contracting the five rank-6 tensors involves contracting from left to right 2 edges at a time, by summing over 2^2 terms four times.

we avoid encountering large rank tensor by contracting the given graph cleverly? The answer is yes. To see this, consider the following two different contraction strategies in Figures 9.5 and 9.6.

Observe that the first strategy in Figure 9.5 has max-rank 5, while the second strategy in Figure 9.6 has max-rank 2; the two strategies differ in their contraction order. It is therefore important to specify a strategic contraction order that yields low max-rank.

There are a number of techniques that can help keeping the rank low [364]. For instance, splitting a tensor into a number of smaller ones (e.g., via singular value decomposition) allows for more degrees of freedom in the contraction. One can also make approximations by dropping some indices regarded as unimportant.

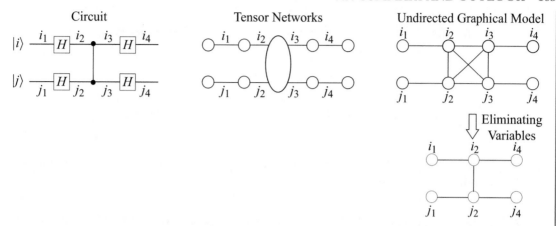

Figure 9.7: Converting from a quantum circuit, to a tensor network, then to an undirected graphical model. Note on bottom-right panel is the reduced graph using a technique called variable elimination.

Undirected Graphical Model

One of the ways to further reduce the cost of contracting a tensor network is by the undirected graphical model [365]. In essence, it combines the ideas from the tensor network and Feynman's sum-over-path approach, as seen from Chapter 3.

We start by examine an example undirected graphical model, derived from a quantum circuit, as shown in Figure 9.7.

Observe that CZ gate is diagonal, suggesting that the rank-4 tensor for CZ gate is redundant. We can further simplify the graph by writing down the path sums for the circuit. Suppose we want to calculate the amplitude for the outcome $|11\rangle$:

$$\langle 11|U|00\rangle = \sum_{i_3,j_3,i_2,j_2} \langle 11|H \otimes H|i_3 j_3\rangle \langle i_3 j_3|CZ|i_2 j_2\rangle \langle i_2 j_2|H \otimes H|00\rangle$$

$$= \sum_{i_2,j_2} \langle 11|H \otimes H|i_2 j_2\rangle \langle i_2 j_2|CZ|i_2 j_2\rangle \langle i_2 j_2|H \otimes H|00\rangle .$$

Note that $\langle i_3 j_3|CZ|i_2 j_2\rangle$ is non-zero if and only if $i_3 j_3 = i_2 j_2$. Therefore, we can simplify the graph by identifying diagonal gates and replace them with the corresponding components in Figure 9.8. This technique, in many cases, can drastically reduce the number of indices we need to sum over, and thus yield a more efficient simulation of the same circuit.

9.5   SUMMARY AND OUTLOOK

Simulators have always played critical roles in systems designs. For classical computers, we often times boostrap from simulators to hardware, and from hardware to larger hardware. Simulation

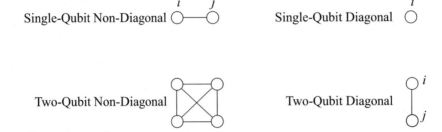

Figure 9.8: In the undirected graphical model, diagonal gates have simplified graph components with fewer indices to sum over.

can model the trajectory of an ideal computation as well as the effects of noise. It remains a fundamental challenge to implement (most likely partial) simulation of quantum processes efficiently and scalably.

Further Reading

Simulating realistic models of noise is non-trivial. Advanced techniques have been developed to balance between simulability and efficiency. For instance, [350, 366] extends noise simulation from using Pauli channels to Clifford channels. [367, 368] apply quasi-probability approximations.

Similarly, for circuit simulations, methods [180, 369] have been developed that extends the Gottesman–Knill theorem to cover more ground (i.e., Clifford and a small number of T gates) at the cost of increased complexity.

CHAPTER 10

Concluding Remarks

The idea of using quantum mechanics, the laws that govern all fundamental particles in the universe, to process information has revolutionized the theory of computing. And soon, nearly half of a century after its first proposal, a practical quantum computer may finally be built. Quantum machines may soon be capable of performing calculations in chemistry, physics, and other fields that are extremely difficult or even impossible for today's conventional computers. Yet a significant gap exists between the theory of quantum algorithms and the devices that will support them. Architects and systems researchers are needed to fill this gap, designing machines and software tools that will efficiently map quantum applications to the constraints of real physical machines.

In this book, we put most emphasis on the design of NISQ computer systems. But it is important not to lose sight of the long-term goal of realizing large-scale fault-tolerant (FT) quantum computers. Optimizing for NISQ systems may appear overwhelming; some even argue that managing N qubits requires precise control over $O(2^N)$ continuous variables (i.e., complex amplitudes). However, that is not the case for the following two reasons. (i) Nature is kind enough to let us vary $O(2^N)$ amplitudes via $O(4N)$ knobs, that is I, X, Y, Z controls for each qubit. The linear number of control parameters is much more manageable than an exponential one. (ii) These knobs can be digital once we are beyond the NISQ era. The lessons we learn for physically driving qubits with analog pulses will pave the way for scalable fault-tolerant systems in the future. The theory of QEC beautifully discretizes noise and protects programs against errors. There is so far no fundamental reason to believe that such FT machines cannot be built.

Quantum Computers can be Digital

It is tempting to view quantum computing as an analog enterprise, with its exponentially complex superposition and probabilistic outcomes of measurement. Remarkably, most quantum computer designs follow a digital discipline (other than the D-Wave quantum annealer, which itself uses a fixed set of control values during its analog annealing process). In particular, this is accomplished through *quantum error correction codes* and the measurement of error syndromes through ancilla qubits (scratch qubits). Imagine a three-qubit quantum majority code in which a logical "0" is encoded as "000" and a logical "1" is encoded as "111." Just as with a classical majority code, a single bit-flip error can be corrected by restoring to the majority value. Unlike a classical code, however, we cannot directly measure the qubits, or their quantum state will be destroyed. Although the errors to the qubits are actually continuous, the effect of measuring the ancilla in syndrome measurements is to discretize the errors, as well as inform us whether an error occurred so that it can be corrected. With this methodology, quantum states are restored in a modular way for even a large quantum computer. Furthermore, operations on error-corrected qubits can be viewed as digital rather than analog, and only a small number of universal operations are needed for universal quantum computation. The standard set consists of the Hadamard gate (H), $\pi/8$-phase gate (T), and controlled-NOT gate (CNOT). Through careful design and engineering, error correction codes and this small set of precise operations will lead to machines that could support practical quantum computation.

It is true that quantum error correction codes have historically required enormous overhead that would be impractical for the foreseeable future. However, this overhead is constantly being reduced through improvements in error-corrected gate methods and reduction in the physical error rates. It is expected that qubit optimized instances of the surface code [22] will demonstrate quantum error correction with 20 qubits in the near term. Surface codes are topological codes that are under-constrained (they also only require near-neighbor 2D qubit connectivity). They are under-constrained in that error syndromes do not uniquely determine the pattern of physical errors that actually occurred. Instead, a maximum likelihood calculation must be computed offline to determine which errors to correct. Intuitively, this under-constrained coding allows more errors to be corrected by fewer physical qubits.

Effective Error-Mitigation Techniques

High physical error rates in quantum devices can lead to high error-correction overhead, even for surface codes. Physical error-mitigation techniques, however, promise to make devices more reliable and make low-overhead error-correction codes possible. These error-mitigation techniques rely on examining the physical basis of the error. For most solid-state systems, the qubit is designed from more primitive physical elements. An active area of research is combining noisy qubits to generate less noisy qubits at the physical level. One example is a proposed four-element superconducting ensemble [370], in which noise continuously transfers from two transmons to two resonant cavities. The noise is then removed from the cavities through a combination of control pulses and dissipation. It is expected that this technique can improve the effective logical qubit lifetimes against photon losses and dephasing error by a factor of more than 40. Another method for physical error mitigation is to use some qubits not for computation but to control the noise source. Trapped-ion machines rely on the shared ion motion to two-qubit gates. A "cooling ion" of a different species can be used to remove noise in the motion to improve the gates between data ions.

In addition to enabling more scalable error-corrected quantum computing, error-mitigation techniques will likely be effective enough to allow small machines of 100–1000 qubits to run some applications without error correction. Additionally, some application-level error-correction is possible. For example, an encoding of quantum chemistry problems, call Generalized Superfast [371], can correct for a single qubit error. This encoding is one of the most efficient discretizations of fermionic quantum simulation, and thus its error-correction properties come at essentially no extra overhead.

Overall, the outlook for quantum computation is promising due to the combination of digital modular design, error-correction, error-mitigation, and application-level fault-tolerance. Hardware continues to scale in performance, where the most recent example is the ion-trap based quantum computer by IonQ demonstrating limited gate operation on up to 79 qubits [56] with fully connected, high fidelity entangling operations on 11 qubits at a time. Additionally, algorithmic and compiler optimizations show signs of orders-of-magnitude reductions in resources required by quantum applications in terms of qubits, operations, and reliability.

Achieving Greater Efficiency by Breaking Abstractions

Practical quantum computation may be achievable in the next few years, but applications will need to be error tolerant and make the best use of a relatively small number of quantum bits and operations. Compilation tools will play a critical role in achieving these goals, but they will have to break traditional abstractions and be customized for machine and device characteristics in a manner never before seen in classical computing. Following the overarching theme of achieving greater efficiency by breaking abstractions, we make the point once again using the following three typical examples from Chapter 6.

- First, compilers can target not only a specific program input and machine size, but the condition of each qubit and link between qubits on a particular day! Several recent quantum circuit mapping and scheduling techniques use heuristics to optimize for specific program input, physical machine size and physical topology. Here, we expose device constraints to the mapper and scheduler in the compiler.

- Second, instead of compiling to an instruction set, compilers can directly target a set of analog control pulses. When we aggregate the instructions, a much more efficient overall pulse can usually be found. In this case, we connect the instruction set architecture with the hardware characteristics.

- Third, instead of using binary logic to target two-level qubits, compilers can target an n-ary logic composed of qudits. By strategically occupying the (generally more noisy) third energy state, we can sometimes significantly shorten critical paths of computation and improve the overall success rate. This is an example where we combine high-level logic (algorithm) design with compiler optimizations.

In Need for a New Systems Stack

Although there already exists an ecosystem of layered quantum software tools and abstractions that serve as an interface between those layers, it is perhaps premature and fallacious to follow a model too similar to classical software. Quantum computing is at a similar stage of development as classical computing in the 1950s. This is actually exciting because there are so many interesting problems to be solved. Specifically, resources are very scarce and we are motivated to break abstractions and pay for efficiency with greater software complexity. How much of what we learn in the next five years will carry forward to a future of much larger quantum machines? Perhaps more than we might think, as it would be hard to imagine a future in which qubits and quantum operations are not costly. Some physical details may always be exposed. Even classical computing is regressing slightly toward less abstraction as device variability increases and the end of Dennard scaling puts pressure on architectures to become more energy efficient.

Quantum computing hardware continues to develop at an impressive pace. Pushing the limits of engineering and technology, these machines will most certainly require software adaptation to hardware constraints such as device variation, operating errors, and environmental noise. To enable practical quantum computation, systems designers will take the responsibility to efficiently map high-level quantum algorithms to resource-constrained quantum machines, which requires optimizations at every layer of the systems stack. This includes the design of (i) a digital modular architecture with feasible error-correction, error-mitigation and application-level fault-tolerance, (ii) an expressive programming language that allows for scalable programming, (iii) an automated compilation and memory management framework that optimizes over algorithm-specific and hardware-specific constraints, (iv) an integrated quantum-classical co-processing scheme that enables efficient execution of hybrid algorithms, and (v) scalable software and hardware verification that gives us confidence on the programs we write and machines we build.

Once we have functional quantum computers, we may even be able to use quantum algorithms to implement theorem provers and constraint solvers. Yet, we will always be bootstrapping from simulator to hardware, from hardware to larger hardware. This is, of course, similar to our experience with classical microprocessors, but perhaps more challenging since each qubit we add to future machines makes the verification and simulation problem exponentially more difficult for current machines.

Bibliography

[1] Tzvetan S. Metodi, Arvin I. Faruque, and Frederic T. Chong. *Quantum Computing for Computer Architects*. Morgan & Claypool Publishers, 2011. DOI: 10.2200/s00066ed1v01y200610cac001 xvi

[2] Paul Benioff. The computer as a physical system: A microscopic quantum mechanical Hamiltonian model of computers as represented by turing machines. *Journal of Statistical Physics*, 22:563–591, May 1980. DOI: 10.1007/bf01011339 3, 16

[3] Charles H. Bennett. Logical reversibility of computation. *IBM J. Res. Dev.*, 17(6):525–532, November 1973. DOI: 10.1147/rd.176.0525 3

[4] Richard P. Feynman. Simulating physics with computers. *International Journal of Theoretical Physics*, 21(6–7):467–488, June 1982. DOI: 10.1201/9780429500459-11 3, 9

[5] Seth Lloyd. Universal quantum simulators. *Science*, pages 1073–1078, 1996. DOI: 10.1126/science.273.5278.1073 3

[6] David Deutsch. Quantum theory, the church-turing principle and the universal quantum computer. *Proc. of the Royal Society of London. A. Mathematical and Physical Sciences*, 400(1818):97–117, 1985. DOI: 10.1098/rspa.1985.0070 4, 63

[7] David Elieser Deutsch. Quantum computational networks. *Proc. of the Royal Society of London. A. Mathematical and Physical Sciences*, 425(1868):73–90, 1989. DOI: 10.1098/rspa.1989.0099 4

[8] David Z. Albert. On quantum-mechanical automata. *Physics Letters A*, 98(5–6):249–252, 1983. DOI: 10.1016/0375-9601(83)90863-0 4

[9] Ethan Bernstein and Umesh Vazirani. Quantum complexity theory. *SIAM Journal on Computing*, 26(5):1411–1473, 1997. DOI: 10.1145/167088.167097 4, 62, 65, 103

[10] Daniel R. Simon. On the power of quantum computation. *SIAM Journal on Computing*, 26(5):1474–1483, 1997. DOI: 10.1137/s0097539796298637 4

[11] Peter W. Shor. Algorithms for quantum computation: Discrete logarithms and factoring. In *Proc. 35th Annual Symposium on Foundations of Computer Science*, pages 124–134, IEEE, 1994. DOI: 10.1109/sfcs.1994.365700 4, 69, 121

[12] Peter W. Shor. Polynomial-time algorithms for prime factorization and discrete logarithms on a quantum computer. *SIAM Review*, 41(2):303–332, 1999. DOI: 10.1137/s0097539795293172 4, 11, 57, 69, 99

[13] Lov K. Grover. A fast quantum mechanical algorithm for database search. *ArXiv Preprint quant-ph/9605043*, 1996. DOI: 10.1145/237814.237866 4, 11, 57, 69, 99, 121

[14] John Preskill. Quantum computing in the NISQ era and beyond. *Quantum*, 2:79, 2018. DOI: 10.22331/q-2018-08-06-79 5, 10, 70, 138

[15] Catherine C. McGeoch. Adiabatic quantum computation and quantum annealing: Theory and practice. *Synthesis Lectures on Quantum Computing*, 5(2):1–93, 2014. DOI: 10.2200/s00585ed1v01y201407qmc008 6

[16] Arnab Das and Bikas K. Chakrabarti. Colloquium: Quantum annealing and analog quantum computation. *Reviews of Modern Physics*, 80(3):1061, 2008. DOI: 10.1103/revmodphys.80.1061 6

[17] Andrew M. Childs, Edward Farhi, and John Preskill. Robustness of adiabatic quantum computation. *Physical Review A*, 65(1):012322, 2001. DOI: 10.1103/physreva.65.012322 6

[18] Dorit Aharonov, Wim Van Dam, Julia Kempe, Zeph Landau, Seth Lloyd, and Oded Regev. Adiabatic quantum computation is equivalent to standard quantum computation. *SIAM Review*, 50(4):755–787, 2008. DOI: 10.1137/080734479 6

[19] Tadashi Kadowaki and Hidetoshi Nishimori. Quantum annealing in the transverse Ising model. *Physical Review E*, 58(5):5355, 1998. DOI: 10.1103/physreve.58.5355 6

[20] Aleta Berk Finnila, M. A. Gomez, C. Sebenik, Catherine Stenson, and Jimmie D. Doll. Quantum annealing: A new method for minimizing multidimensional functions. *Chemical Physics Letters*, 219(5–6):343–348, 1994. DOI: 10.1016/0009-2614(94)00117-0 6

[21] Mohammad H. S. Amin, Dmitri V. Averin, and James A. Nesteroff. Decoherence in adiabatic quantum computation. *Physical Review A*, 79(2):022107, 2009. DOI: 10.1103/physreva.79.022107 6

[22] Eric Dennis, Alexei Kitaev, Andrew Landahl, and John Preskill. Topological quantum memory. *Journal of Mathematical Physics*, 43(9):4452–4505, 2002. DOI: 10.1063/1.1499754 6, 77, 141, 166

[23] Daniel Gottesman. An introduction to quantum error correction and fault-tolerant quantum computation. In *Quantum Information Science and its Contributions to Mathematics, Proceedings of Symposia in Applied Mathematics*, 68:13–58, 2010. DOI: 10.1090/psapm/068/2762145 6, 150

[24] Simon J. Devitt, William J. Munro, and Kae Nemoto. Quantum error correction for beginners. *Reports on Progress in Physics*, 76(7):076001, 2013. DOI: 10.1088/0034-4885/76/7/076001 6

[25] Barbara M. Terhal. Quantum error correction for quantum memories. *Reviews of Modern Physics*, 87(2):307, 2015. DOI: 10.1103/revmodphys.87.307 6, 141, 150

[26] Chetan Nayak, Steven H. Simon, Ady Stern, Michael Freedman, and Sankar Das Sarma. Non-abelian anyons and topological quantum computation. *Reviews of Modern Physics*, 80(3):1083, 2008. DOI: 10.1103/revmodphys.80.1083 6

[27] Austin G. Fowler, Matteo Mariantoni, John M. Martinis, and Andrew N. Cleland. Surface codes: Towards practical large-scale quantum computation. *Physical Review A*, 86(3):032324, 2012. DOI: 10.1103/physreva.86.032324 6, 77, 117

[28] Sergey Bravyi and Alexei Kitaev. Universal quantum computation with ideal Clifford gates and noisy ancillas. *Physical Review A*, 71(2):022316, 2005. DOI: 10.1103/physreva.71.022316 6, 148

[29] Robert Raussendorf, Daniel E. Browne, and Hans J. Briegel. Measurement-based quantum computation on cluster states. *Physical Review A*, 68(2):022312, 2003. DOI: 10.1103/physreva.68.022312 6

[30] Michael A. Nielsen. Cluster-state quantum computation. *Reports on Mathematical Physics*, 57(1):147–161, 2006. DOI: 10.1016/s0034-4877(06)80014-5 6

[31] Frederic T. Chong, Diana Franklin, and Margaret Martonosi. Programming languages and compiler design for realistic quantum hardware. *Nature*, 549(7671):180, 2017. DOI: 10.1038/nature23459 8, 12, 124, 125

[32] Margaret Martonosi and Martin Roetteler. Next steps in quantum computing: Computer science's role. *ArXiv Preprint ArXiv:1903.10541*, 2019. 8, 12

[33] Mathias Soeken, Thomas Haener, and Martin Roetteler. Programming quantum computers using design automation. In *Design, Automation and Test in Europe Conference and Exhibition (DATE)*, pages 137–146, IEEE, 2018. DOI: 10.23919/date.2018.8341993 9

[34] Nicolas Gisin and Rob Thew. Quantum communication. *Nature Photonics*, 1(3):165, 2007. DOI: 10.1038/nphoton.2007.22 9

[35] H.-J. Briegel, Wolfgang Dür, Juan I. Cirac, and Peter Zoller. Quantum repeaters: The role of imperfect local operations in quantum communication. *Physical Review Letters*, 81(26):5932, 1998. DOI: 10.1103/physrevlett.81.5932 9

[36] Sreraman Muralidharan, Jungsang Kim, Norbert Lütkenhaus, Mikhail D. Lukin, and Liang Jiang. Ultrafast and fault-tolerant quantum communication across long distances. *Physical Review Letters*, 112(25):250501, 2014. DOI: 10.1103/PhysRevLett.112.250501 9

[37] Romain Alléaume, Cyril Branciard, Jan Bouda, Thierry Debuisschert, Mehrdad Dianati, Nicolas Gisin, Mark Godfrey, Philippe Grangier, Thomas Länger, Norbert Lütkenhaus, et al. Using quantum key distribution for cryptographic purposes: A survey. *Theoretical Computer Science*, 560:62–81, 2014. DOI: 10.1016/j.tcs.2014.09.018 9

[38] Juan Yin, Yuan Cao, Yu-Huai Li, Sheng-Kai Liao, Liang Zhang, Ji-Gang Ren, Wen-Qi Cai, Wei-Yue Liu, Bo Li, Hui Dai, et al. Satellite-based entanglement distribution over 1,200 kilometers. *Science*, 356(6343):1140–1144, 2017. DOI: 10.1126/science.aan3211 9

[39] Boris Korzh, Charles Ci Wen Lim, Raphael Houlmann, Nicolas Gisin, Ming Jun Li, Daniel Nolan, Bruno Sanguinetti, Rob Thew, and Hugo Zbinden. Provably secure and practical quantum key distribution over 307 km of optical fibre. *Nature Photonics*, 9(3):163, 2015. DOI: 10.1038/nphoton.2014.327 9

[40] Robert B. Laughlin and David Pines. The theory of everything. *Proc. of the National Academy of Sciences*, 97(1):28–31, 2000. DOI: 10.7551/mit-press/9780262026215.003.0017 9

[41] Markus Reiher, Nathan Wiebe, Krysta M. Svore, Dave Wecker, and Matthias Troyer. Elucidating reaction mechanisms on quantum computers. *Proc. of the National Academy of Sciences*, 114(29):7555–7560, 2017. DOI: 10.1073/pnas.1619152114 9

[42] Attila Szabo and Neil S. Ostlund. *Modern Quantum Chemistry: Introduction to Advanced Electronic Structure Theory*. Courier Corporation, 2012. 9

[43] Ben P. Lanyon, Cornelius Hempel, Daniel Nigg, Markus Müller, Rene Gerritsma, F. Zähringer, Philipp Schindler, Julio T. Barreiro, Markus Rambach, Gerhard Kirchmair, et al. Universal digital quantum simulation with trapped ions. *Science*, 334(6052):57–61, 2011. DOI: 10.1126/science.1208001 9

[44] Benjamin P. Lanyon, James D. Whitfield, Geoff G. Gillett, Michael E. Goggin, Marcelo P. Almeida, Ivan Kassal, Jacob D. Biamonte, Masoud Mohseni, Ben J. Powell, Marco Barbieri, et al. Towards quantum chemistry on a quantum computer. *Nature Chemistry*, 2(2):106, 2010. DOI: 10.1038/nchem.483 9

[45] Andrew A. Houck, Hakan E. Türeci, and Jens Koch. On-chip quantum simulation with superconducting circuits. *Nature Physics*, 8(4):292, 2012. DOI: 10.1038/nphys2251 9

[46] Peter J. J. O'Malley, Ryan Babbush, Ian D. Kivlichan, Jonathan Romero, Jarrod R. Mc-Clean, Rami Barends, Julian Kelly, Pedram Roushan, Andrew Tranter, Nan Ding, et al. Scalable quantum simulation of molecular energies. *Physical Review X*, 6(3):031007, 2016. DOI: 10.1103/physrevx.6.031007 9, 67, 70

[47] Christian L. Degen, F. Reinhard, and P. Cappellaro. Quantum sensing. *Reviews of Modern Physics*, 89(3):035002, 2017. DOI: 10.1103/revmodphys.89.035002 9

[48] Jens M. Boss, K. S. Cujia, Jonathan Zopes, and Christian L. Degen. Quantum sensing with arbitrary frequency resolution. *Science*, 356(6340):837–840, 2017. DOI: 10.1126/science.aam7009 9

[49] Sebastian Zaiser, Torsten Rendler, Ingmar Jakobi, Thomas Wolf, Sang-Yun Lee, Samuel Wagner, Ville Bergholm, Thomas Schulte-Herbrüggen, Philipp Neumann, and Jörg Wrachtrup. Enhancing quantum sensing sensitivity by a quantum memory. *Nature Communications*, 7:12279, 2016. DOI: 10.1038/ncomms12279 9

[50] Robert R. Schaller. Moore's law: Past, present and future. *IEEE Spectrum*, 34(6):52–59, 1997. DOI: 10.1109/6.591665 10, 15

[51] IBM Unveils World's First Integrated Quantum Computing System for Commercial Use. https://newsroom.ibm.com/2019-01-08-IBM-Unveils-Worlds-First-Integrated-Quantum-Computing-System-for-Commercial-Use 10, 77, 119, 138, 140

[52] IBM Opens Quantum Computation Center in New York. https://newsroom.ibm.com/2019-09-18-IBM-Opens-Quantum-Computation-Center-in-New-York-Brings-Worlds-Largest-Fleet-of-Quantum-Computing-Systems-Online-Unveils-New-53-Qubit-Quantum-System-for-Broad-Use 10

[53] Quantum Supremacy Using a Programmable Superconducting Processor. https://ai.googleblog.com/2019/10/quantum-supremacy-using-programmable.html 10

[54] The Future of Quantum Computing is Counted in Qubits. https://newsroom.intel.com/news/future-quantum-computing-counted-qubits/#gs.qih7ym 10

[55] Intel Introduces "Horse Ridge" to Enable Commercially Viable Quantum Computers. https://newsroom.intel.com/news/intel-introduces-horse-ridge-enable-commercially-viable-quantum-computers/#gs.qihieg 10

[56] IonQ Newsletter. https://ionq.co/news/december-11-2018 10, 77, 138, 167

[57] Hartmut Häffner, Wolfgang Hänsel, C. F. Roos, Jan Benhelm, Michael Chwalla, Timo Körber, U. D. Rapol, Mark Riebe, P. O. Schmidt, Christoph Becher, et al. Scalable multiparticle entanglement of trapped ions. *Nature*, 438(7068):643–646, 2005. DOI: 10.1038/nature04279 11

[58] Dietrich Leibfried, Brian DeMarco, Volker Meyer, David Lucas, Murray Barrett, Joe Britton, Wayne M. Itano, B. Jelenković, Chris Langer, Till Rosenband, et al. Experimental demonstration of a robust, high-fidelity geometric two ion-qubit phase gate. *Nature*, 422(6930):412–415, 2003. DOI: 10.1038/nature01492 11

[59] Ferdinand Schmidt-Kaler, Hartmut Häffner, Mark Riebe, Stephan Gulde, Gavin P. T. Lancaster, Thomas Deuschle, Christoph Becher, Christian F. Roos, Jürgen Eschner, and Rainer Blatt. Realization of the Cirac–Zoller controlled-not quantum gate. *Nature*, 422(6930):408–411, 2003. DOI: 10.1038/nature01494 11

[60] Matthias Steffen, M. Ansmann, Radoslaw C. Bialczak, Nadav Katz, Erik Lucero, R. McDermott, Matthew Neeley, Eva Maria Weig, Andrew N. Cleland, and John M. Martinis. Measurement of the entanglement of two superconducting qubits via state tomography. *Science*, 313(5792):1423–1425, 2006. DOI: 10.1126/science.1130886 11, 51

[61] Leonardo DiCarlo, Jerry M. Chow, Jay M. Gambetta, Lev S. Bishop, Blake R. Johnson, D. I. Schuster, J. Majer, Alexandre Blais, Luigi Frunzio, S. M. Girvin, et al. Demonstration of two-qubit algorithms with a superconducting quantum processor. *Nature*, 460(7252):240–244, 2009. DOI: 10.1038/nature08121 11, 49, 51

[62] Jerry M. Chow, Jay M. Gambetta, A. D. Córcoles, Seth T. Merkel, John A. Smolin, Chad Rigetti, S. Poletto, George A. Keefe, Mary B. Rothwell, J. R. Rozen, et al. Universal quantum gate set approaching fault-tolerant thresholds with superconducting qubits. *Physical Review Letters*, 109(6):060501, 2012. DOI: 10.1103/physrevlett.109.060501 11

[63] Sarah Sheldon, Easwar Magesan, Jerry M. Chow, and Jay M. Gambetta. Procedure for systematically tuning up cross-talk in the cross-resonance gate. *Physical Review A*, 93(6):060302, 2016. DOI: 10.1103/physreva.93.060302 11

[64] John P. Gaebler, Ting Rei Tan, Y. Lin, Y. Wan, R. Bowler, Adam C. Keith, S. Glancy, K. Coakley, E. Knill, D. Leibfried, et al. High-fidelity universal gate set for be 9+ ion qubits. *Physical Review Letters*, 117(6):060505, 2016. DOI: 10.1103/physrevlett.117.060505 11

[65] K. Wright, K. M. Beck, S. Debnath, J. M. Amini, Y. Nam, N. Grzesiak, J. S. Chen, N. C. Pisenti, M. Chmielewski, C. Collins, K. M. Hudek, J. Mizrahi, J. D. Wong-Campos, S. Allen, J. Apisdorf, P. Solomon, M. Williams, A. M. Ducore, A. Blinov, S. M. Kreikemeier, V. Chaplin, M. Keesan, C. Monroe, and J. Kim. Benchmarking an 11-qubit quantum computer. *Nature Communications*, 10(1):5464, 2019. DOI: 10.1038/s41467-019-13534-2 11

[66] Frank Arute, Kunal Arya, Ryan Babbush, Dave Bacon, Joseph C. Bardin, Rami Barends, Rupak Biswas, Sergio Boixo, Fernando G. S. L. Brandao, David A. Buell, et al. Quantum supremacy using a programmable superconducting processor. *Nature*, 574(7779):505–510, 2019. DOI: 10.1038/s41586-019-1666-5 11, 140

[67] J. M. Pino, J. M. Dreiling, C. Figgatt, J. P. Gaebler, S. A. Moses, C. H. Baldwin, M. Foss-Feig, D. Hayes, K. Mayer, C. Ryan-Anderson, et al. Demonstration of the QCCD trapped-ion quantum computer architecture. *ArXiv Preprint ArXiv:2003.01293*, 2020. 11

[68] Edward Farhi, Jeffrey Goldstone, and Sam Gutmann. A quantum approximate optimization algorithm. *ArXiv Preprint ArXiv:1411.4028*, 2014. 11, 67, 68, 70

[69] Dmitri Maslov. Basic circuit compilation techniques for an ion-trap quantum machine. *New Journal of Physics*, 19(2):023035, 2017. DOI: 10.1088/1367-2630/aa5e47 12

[70] National Academies of Sciences, Engineering and Medicine, *Quantum computing: Progress and prospects*. National Academies Press, 2019. DOI: 10.17226/25196 12

[71] Antonio Acín, Immanuel Bloch, Harry Buhrman, Tommaso Calarco, Christopher Eichler, Jens Eisert, Daniel Esteve, Nicolas Gisin, Steffen J. Glaser, Fedor Jelezko, et al. The quantum technologies roadmap: A European community view. *New Journal of Physics*, 20(8):080201, 2018. DOI: 10.1088/1367-2630/aad1ea 12

[72] Andrew Waterman, Yunsup Lee, David A. Patterson, and Krste Asanovic. The RISC-V instruction set manual, volume I: Base user-level ISA. *EECS Department, Tech. Rep. UCB/EECS-2011-62*, 116, UC Berkeley, 2011. DOI: 10.21236/ada605735 15

[73] Rolf Landauer. Irreversibility and heat generation in the computing process. *IBM Journal of Research and Development*, 5(3):183–191, 1961. DOI: 10.1147/rd.441.0261 16

[74] Charles H. Bennett. Logical reversibility of computation. *IBM Journal of Research and Development*, 17(6):525–532, 1973. DOI: 10.1147/rd.176.0525 16, 120

[75] Albert Einstein, Boris Podolsky, and Nathan Rosen. Can quantum-mechanical description of physical reality be considered complete? *Physical Review*, 47(10):777, 1935. DOI: 10.1007/978-3-322-91080-6_6 21, 23

[76] John S. Bell. On the Einstein Podolsky rosen paradox. *Physics Physique Fizika*, 1(3):195, 1964. DOI: 10.1142/9789812386540_0002 21, 23, 24

[77] Richard Phillips Feynman. Space-time approach to non-relativistic quantum mechanics. In *Feynman's Thesis—A New Approach To Quantum Theory*, pages 71–109. World Scientific, 2005. DOI: 10.1103/revmodphys.20.367 22

[78] Richard P. Feynman, Albert R. Hibbs, and Daniel F. Styer. *Quantum Mechanics and Path Integrals*. Courier Corporation, 2010. DOI: 10.1063/1.3048320 22

[79] Scott Aaronson. Bell inequality violation finally done right. https://www.scottaaronson.com/blog/?p=2464 24

[80] John F. Clauser, Michael A. Horne, Abner Shimony, and Richard A. Holt. Proposed experiment to test local hidden-variable theories. *Physical Review Letters*, 23(15):880, 1969. DOI: 10.1103/physrevlett.24.549 24

[81] Ryszard Horodecki, Paweł Horodecki, Michał Horodecki, and Karol Horodecki. Quantum entanglement. *Reviews of Modern Physics*, 81(2):865, 2009. DOI: 10.1103/revmodphys.81.865 24

[82] William K. Wootters and Wojciech H. Zurek. A single quantum cannot be cloned. *Nature*, 299(5886):802–803, 1982. DOI: 10.1038/299802a0 25

[83] Daniel Oliveira, Laércio Pilla, Nathan DeBardeleben, Sean Blanchard, Heather Quinn, Israel Koren, Philippe Navaux, and Paolo Rech. Experimental and analytical study of xeon phi reliability. In *Proc. of the International Conference for High Performance Computing, Networking, Storage and Analysis*, page 28, ACM, 2017. DOI: 10.1145/3126908.3126960 26

[84] IBM Quantum Computing. https://www.ibm.com/quantum-computing/ 26, 83

[85] Karl Kraus, Arno Böhm, John D. Dollard, and W. H. Wootters. States, effects, and operations: Fundamental notions of quantum theory. Lectures in mathematical physics, University of Texas, Austin, TX. *Lecture Notes in Physics*, 190, 1983. DOI: 10.1007/3-540-12732-1 41

[86] Michael A. Nielsen and Isaac Chuang. Quantum computation and quantum information, Cambridge University Press, 2002. DOI: 10.1017/cbo9780511976667 42, 94, 95, 96, 141, 146, 150

[87] David P. DiVincenzo. The physical implementation of quantum computation. *Fortschritte der Physik: Progress of Physics*, 48(9–11):771–783, 2000. DOI: 10.1002/3527603182.ch1 43

[88] Steve Olmschenk, Kelly C. Younge, David L. Moehring, Dzmitry N. Matsukevich, Peter Maunz, and Christopher Monroe. Manipulation and detection of a trapped yb+ hyperfine qubit. *Physical Review A*, 76(5):052314, 2007. DOI: 10.1103/physreva.76.052314 44

[89] Boris B. Blinov, Dietrich Leibfried, C. Monroe, and David J. Wineland. Quantum computing with trapped ion hyperfine qubits. *Quantum Information Processing*, 3(1–5):45–59, 2004. DOI: 10.1007/0-387-27732-3_4 44

[90] Rachel Noek, Geert Vrijsen, Daniel Gaultney, Emily Mount, Taehyun Kim, Peter Maunz, and Jungsang Kim. High speed, high fidelity detection of an atomic hyperfine qubit. *Optics Letters*, 38(22):4735–4738, 2013. DOI: 10.1364/ol.38.004735 44

[91] A. H. Myerson, D. J. Szwer, S. C. Webster, D. T. C. Allcock, M. J. Curtis, G. Imreh, J. A. Sherman, D. N. Stacey, A. M. Steane, and D. M. Lucas. High-fidelity readout of trapped-ion qubits. *Physical Review Letters*, 100(20):200502, 2008. DOI: 10.1103/physrevlett.100.200502 44

[92] Juan I. Cirac and Peter Zoller. Quantum computations with cold trapped ions. *Physical Review Letters*, 74(20):4091, 1995. DOI: 10.1103/physrevlett.74.4091 45

[93] Anders Sørensen and Klaus Mølmer. Quantum computation with ions in thermal motion. *Physical Review Letters*, 82(9):1971, 1999. DOI: 10.1103/physrevlett.82.1971 45

[94] E. Solano, R. L. de Matos Filho, and N. Zagury. Deterministic bell states and measurement of the motional state of two trapped ions. *Physical Review A*, 59(4):R2539, 1999. DOI: 10.1103/physreva.59.r2539 45

[95] G. J. Milburn, S. Schneider, and D. F. V. James. Ion trap quantum computing with warm ions. *Fortschritte der Physik: Progress of Physics*, 48(9–11):801–810, 2000. DOI: 10.1002/3527603182.ch3 45

[96] Wolfgang Paul. Electromagnetic traps for charged and neutral particles. *Reviews of Modern Physics*, 62(3):531, 1990. DOI: 10.1103/revmodphys.62.531 47

[97] Hans G. Dehmelt. Radiofrequency spectroscopy of stored ions I: Storage. In *Advances in Atomic and Molecular Physics*, 3:53–72, Elsevier, 1968. DOI: 10.1016/s0065-2199(08)60170-0 47

[98] Christopher Monroe and Jungsang Kim. Scaling the ion trap quantum processor. *Science*, 339(6124):1164–1169, 2013. DOI: 10.1126/science.1231298 48

[99] G.-D. Lin, S.-L. Zhu, Rajibul Islam, Kihwan Kim, M.-S. Chang, Simcha Korenblit, Christopher Monroe, and L.-M. Duan. Large-scale quantum computation in an anharmonic linear ion trap. *EPL (Europhysics Letters)*, 86(6):60004, 2009. DOI: 10.1209/0295-5075/86/60004 48

[100] Peter Lukas Wilhelm Maunz. Characterization of two-qubit quantum gates in Sandia's high optical access surface ion trap. *Technical Report*, Sandia National Lab. (SNL-NM), Albuquerque, NM, 2016. 48

[101] Brian David Josephson. Possible new effects in superconductive tunnelling. *Physics Letters*, 1(7):251–253, 1962. DOI: 10.1016/0031-9163(62)91369-0 48

[102] B. D. Josephson. Coupled superconductors. *Reviews of Modern Physics*, 36(1):216, 1964. DOI: 10.1103/revmodphys.36.216 48

[103] Yu Nakamura, Yu A. Pashkin, and Jaw Shen Tsai. Coherent control of macroscopic quantum states in a single-cooper-pair box. *Nature*, 398(6730):786–788, 1999. DOI: 10.1038/19718 49

[104] Denis Vion, A. Aassime, Audrey Cottet, P. L. Joyez, H. Pothier, C. Urbina, Daniel Esteve, and Michel H. Devoret. Manipulating the quantum state of an electrical circuit. *Science*, 296(5569):886–889, 2002. DOI: 10.1126/science.1069372 49

[105] Tim Duty, D. Gunnarsson, K. Bladh, and Per Delsing. Coherent dynamics of a Josephson charge qubit. *Physical Review B*, 69(14):140503, 2004. DOI: 10.1103/physrevb.69.140503 49

[106] Jens Koch, M. Yu Terri, Jay Gambetta, Andrew A. Houck, D. I. Schuster, J. Majer, Alexandre Blais, Michel H. Devoret, Steven M. Girvin, and Robert J. Schoelkopf. Charge-insensitive qubit design derived from the cooper pair box. *Physical Review A*, 76(4):042319, 2007. DOI: 10.1103/physreva.76.042319 49

[107] T. P. Orlando, J. E. Mooij, Lin Tian, Caspar H. Van Der Wal, L. S. Levitov, Seth Lloyd, and J. J. Mazo. Superconducting persistent-current qubit. *Physical Review B*, 60(22):15398, 1999. DOI: 10.1103/physrevb.60.15398 49

[108] J. E. Mooij, T. P. Orlando, L. Levitov, Lin Tian, Caspar H. Van der Wal, and Seth Lloyd. Josephson persistent-current qubit. *Science*, 285(5430):1036–1039, 1999. DOI: 10.1126/science.285.5430.1036 49

[109] Michael Tinkham. *Introduction to Superconductivity*. Courier Corporation, 2004. DOI: 10.1063/1.2807811 49

[110] Ioan M. Pop, Kurtis Geerlings, Gianluigi Catelani, Robert J. Schoelkopf, Leonid I. Glazman, and Michel H. Devoret. Coherent suppression of electromagnetic dissipation due to superconducting quasiparticles. *Nature*, 508(7496):369, 2014. DOI: 10.1038/nature13017 50

[111] John M. Martinis, S. Nam, J. Aumentado, and C. Urbina. Rabi oscillations in a large Josephson-junction qubit. *Physical Review Letters*, 89(11):117901, 2002. DOI: 10.1103/physrevlett.89.117901 50

[112] Raymond W. Simmonds, K. M. Lang, Dustin A. Hite, S. Nam, David P. Pappas, and John M. Martinis. Decoherence in Josephson phase qubits from junction resonators. *Physical Review Letters*, 93(7):077003, 2004. DOI: 10.1103/physrevlett.93.077003 50

[113] Philip Krantz, Morten Kjaergaard, Fei Yan, Terry P. Orlando, Simon Gustavsson, and William D. Oliver. A quantum engineer's guide to superconducting qubits. *Applied Physics Reviews*, 6(2):021318, 2019. DOI: 10.1063/1.5089550 50, 51

[114] Daniel Thomas Sank. Fast, accurate state measurement in superconducting qubits. Ph.D. thesis, UC Santa Barbara, 2014. 50

[115] Uri Vool and Michel Devoret. Introduction to quantum electromagnetic circuits. *International Journal of Circuit Theory and Applications*, 45(7):897–934, 2017. DOI: 10.1002/cta.2359 50

[116] Edwin T. Jaynes and Frederick W. Cummings. Comparison of quantum and semiclassical radiation theories with application to the beam maser. *Proc. of the IEEE*, 51(1):89–109, 1963. DOI: 10.1109/proc.1963.1664 50

[117] Alexandre Blais, Ren-Shou Huang, Andreas Wallraff, Steven M. Girvin, and R. Jun Schoelkopf. Cavity quantum electrodynamics for superconducting electrical circuits: An architecture for quantum computation. *Physical Review A*, 69(6):062320, 2004. DOI: 10.1103/physreva.69.062320 50

[118] Vladimir B. Braginsky and F. Ya Khalili. Quantum nondemolition measurements: The route from toys to tools. *Reviews of Modern Physics*, 68(1):1, 1996. DOI: 10.1103/revmodphys.68.1 50

[119] M. D. Reed, L. DiCarlo, B. R. Johnson, L. Sun, D. I. Schuster, L. Frunzio, and R. J. Schoelkopf. High-fidelity readout in circuit quantum electrodynamics using the Jaynes-cummings nonlinearity. *Physical Review Letters*, 105(17):173601, 2010. DOI: 10.1103/physrevlett.105.173601 50

[120] David C. McKay, Christopher J. Wood, Sarah Sheldon, Jerry M. Chow, and Jay M. Gambetta. Efficient z gates for quantum computing. *Physical Review A*, 96(2):022330, 2017. DOI: 10.1103/physreva.96.022330 51

[121] J. Majer, J. M. Chow, J. M. Gambetta, Jens Koch, B. R. Johnson, J. A. Schreier, L. Frunzio, D. I. Schuster, Andrew Addison Houck, Andreas Wallraff, et al. Coupling superconducting qubits via a cavity bus. *Nature*, 449(7161):443, 2007. DOI: 10.1038/nature06184 51

[122] Göran Wendin and V. S. Shumeiko. Superconducting quantum circuits, qubits and computing. *ArXiv Preprint cond-mat/0508729*, 2005. 51

[123] Radoslaw C. Bialczak, Markus Ansmann, Max Hofheinz, Erik Lucero, Matthew Neeley, A. D. O'Connell, Daniel Sank, Haohua Wang, James Wenner, Matthias Steffen, et al. Quantum process tomography of a universal entangling gate implemented with Josephson phase qubits. *Nature Physics*, 6(6):409–413, 2010. DOI: 10.1038/nphys1639 51

[124] Matthew Neeley, Radoslaw C. Bialczak, M. Lenander, Erik Lucero, Matteo Mariantoni, A. D. O'connell, D. Sank, H. Wang, M. Weides, J. Wenner, et al. Generation of three-qubit entangled states using superconducting phase qubits. *Nature*, 467(7315):570, 2010. DOI: 10.1038/nature09418 51

[125] Frederick W. Strauch, Philip R. Johnson, Alex J. Dragt, C. J. Lobb, J. R. Anderson, and F. C. Wellstood. Quantum logic gates for coupled superconducting phase qubits. *Physical Review Letters*, 91(16):167005, 2003. DOI: 10.1103/physrevlett.91.167005 51

[126] G. S. Paraoanu. Microwave-induced coupling of superconducting qubits. *Physical Review B*, 74(14):140504, 2006. DOI: 10.1103/physrevb.74.140504 51

[127] Chad Rigetti and Michel Devoret. Fully microwave-tunable universal gates in superconducting qubits with linear couplings and fixed transition frequencies. *Physical Review B*, 81(13):134507, 2010. DOI: 10.1103/physrevb.81.134507 51

[128] P. C. De Groot, J. Lisenfeld, R. N. Schouten, S. Ashhab, A. Lupaşcu, C. J. P. M. Harmans, and J. E. Mooij. Selective darkening of degenerate transitions demonstrated with two superconducting quantum bits. *Nature Physics*, 6(10):763, 2010. DOI: 10.1038/nphys1733 51

[129] Jerry M. Chow, Jay M. Gambetta, Andrew W. Cross, Seth T. Merkel, Chad Rigetti, and M. Steffen. Microwave-activated conditional-phase gate for superconducting qubits. *New Journal of Physics*, 15(11):115012, 2013. DOI: 10.1088/1367-2630/15/11/115012 51, 53

[130] Daniel Loss and David P. DiVincenzo. Quantum computation with quantum dots. *Physical Review A*, 57(1):120, 1998. DOI: 10.1103/physreva.57.120 54

[131] Bruce E. Kane. A silicon-based nuclear spin quantum computer. *Nature*, 393(6681):133, 1998. DOI: 10.1038/30156 54, 132

[132] Rutger Vrijen, Eli Yablonovitch, Kang Wang, Hong Wen Jiang, Alex Balandin, Vwani Roychowdhury, Tal Mor, and David DiVincenzo. Electron-spin-resonance transistors for quantum computing in silicon-germanium heterostructures. *Physical Review A*, 62(1):012306, 2000. DOI: 10.1103/physreva.62.012306 54

[133] L. C. L. Hollenberg, A. D. Greentree, A. G. Fowler, and C. J. Wellard. Two-dimensional architectures for donor-based quantum computing. *Physical Review B*, 74(4):045311, 2006. DOI: 10.1103/physrevb.74.045311 54

[134] Emanuel Knill, Raymond Laflamme, and Gerald J. Milburn. A scheme for efficient quantum computation with linear optics. *Nature*, 409(6816):46, 2001. DOI: 10.1038/35051009 54

[135] T. B. Pittman, B. C. Jacobs, and J. D. Franson. Probabilistic quantum logic operations using polarizing beam splitters. *Physical Review A*, 64(6):062311, 2001. DOI: 10.1103/physreva.64.062311 54

[136] James D. Franson, M. M. Donegan, M. J. Fitch, B. C. Jacobs, and T. B. Pittman. High-fidelity quantum logic operations using linear optical elements. *Physical Review Letters*, 89(13):137901, 2002. DOI: 10.1103/physrevlett.89.137901 54

[137] A. Yu Kitaev. Fault-tolerant quantum computation by anyons. *Annals of Physics*, 303(1):2–30, 2003. DOI: 10.1016/s0003-4916(02)00018-0 54

[138] M. T. Deng, S. Vaitiekėnas, Esben Bork Hansen, Jeroen Danon, M. Leijnse, Karsten Flensberg, Jesper Nygård, P. Krogstrup, and Charles M. Marcus. Majorana bound state in a coupled quantum-dot hybrid-nanowire system. *Science*, 354(6319):1557–1562, 2016. DOI: 10.1126/science.aaf3961 54

[139] Torsten Karzig, Christina Knapp, Roman M. Lutchyn, Parsa Bonderson, Matthew B. Hastings, Chetan Nayak, Jason Alicea, Karsten Flensberg, Stephan Plugge, Yuval Oreg, et al. Scalable designs for quasiparticle-poisoning-protected topological quantum computation with Majorana zero modes. *Physical Review B*, 95(23):235305, 2017. DOI: 10.1103/physrevb.95.235305 54

[140] R. M. T. Lutchyn, Epam Bakkers, Leo P. Kouwenhoven, Peter Krogstrup, C. M. Marcus, and Y. Oreg. Majorana zero modes in superconductor-semiconductor heterostructures. *Nature Reviews Materials*, 3(5):52–68, 2018. DOI: 10.1038/s41578-018-0003-1 54

[141] Gilles Brassard, Peter Hoyer, and Alain Tapp. Quantum algorithm for the collision problem. *ArXiv Preprint quant-ph/9705002*, 1997. DOI: 10.1007/978-3-642-27848-8_304-2 57

[142] Scott Aaronson and Yaoyun Shi. Quantum lower bounds for the collision and the element distinctness problems. *Journal of the ACM (JACM)*, 51(4):595–605, 2004. DOI: 10.1145/1008731.1008735 57

[143] David Deutsch and Richard Jozsa. Rapid solution of problems by quantum computation. *Proc. of the Royal Society of London. Series A: Mathematical and Physical Sciences*, 439(1907):553–558, 1992. DOI: 10.1098/rspa.1992.0167 62, 64

[144] Alberto Peruzzo, Jarrod McClean, Peter Shadbolt, Man-Hong Yung, Xiao-Qi Zhou, Peter J. Love, Alán Aspuru-Guzik, and Jeremy L. O'brien. A variational eigenvalue solver on a photonic quantum processor. *Nature Communications*, 5:4213, 2014. DOI: 10.1038/ncomms5213 67, 70

[145] Jarrod R. McClean, Jonathan Romero, Ryan Babbush, and Alán Aspuru-Guzik. The theory of variational hybrid quantum-classical algorithms. *New Journal of Physics*, 18(2):023023, 2016. DOI: 10.1088/1367-2630/18/2/023023 67, 70, 75

[146] Aram W. Harrow and Ashley Montanaro. Quantum computational supremacy. *Nature*, 549(7671):203, 2017. DOI: 10.1038/nature23458 70, 85

[147] Ewin Tang. A quantum-inspired classical algorithm for recommendation systems. In *Proc. of the 51st Annual ACM SIGACT Symposium on Theory of Computing*, pages 217–228, 2019. DOI: 10.1145/3313276.3316310 70

[148] Dave Wecker, Matthew B. Hastings, and Matthias Troyer. Progress towards practical quantum variational algorithms. *Physical Review A*, 92(4):042303, 2015. DOI: 10.1103/physreva.92.042303 70

[149] Jonathan Romero, Jonathan P. Olson, and Alan Aspuru-Guzik. Quantum autoencoders for efficient compression of quantum data. *Quantum Science and Technology*, 2(4):045001, 2017. DOI: 10.1088/2058-9565/aa8072 70

[150] Guillaume Verdon, Michael Broughton, and Jacob Biamonte. A quantum algorithm to train neural networks using low-depth circuits. *ArXiv Preprint ArXiv:1712.05304*, 2017. 70

[151] Marcello Benedetti, Delfina Garcia-Pintos, Oscar Perdomo, Vicente Leyton-Ortega, Yunseong Nam, and Alejandro Perdomo-Ortiz. A generative modeling approach for benchmarking and training shallow quantum circuits. *NPJ Quantum Information*, 5(1):45, 2019. DOI: 10.1038/s41534-019-0157-8 70

[152] Scott Aaronson and Alex Arkhipov. The computational complexity of linear optics. In *Proc. of the 43rd Annual ACM Symposium on Theory of Computing*, pages 333–342, ACM, 2011. DOI: 10.1364/qim.2014.qth1a.2 70

[153] Jacques Carolan, Christopher Harrold, Chris Sparrow, Enrique Martín-López, Nicholas J. Russell, Joshua W. Silverstone, Peter J. Shadbolt, Nobuyuki Matsuda, Manabu Oguma, Mikitaka Itoh, et al. Universal linear optics. *Science*, 349(6249):711–716, 2015. DOI: 10.1126/science.aab3642 70

[154] Peter Clifford and Raphaël Clifford. The classical complexity of boson sampling. In *Proc. of the 29th Annual ACM-SIAM Symposium on Discrete Algorithms*, pages 146–155. Society for Industrial and Applied Mathematics, 2018. DOI: 10.1137/1.9781611975031.10 70

[155] Sergio Boixo, Sergei V. Isakov, Vadim N. Smelyanskiy, Ryan Babbush, Nan Ding, Zhang Jiang, Michael J. Bremner, John M. Martinis, and Hartmut Neven. Characterizing quantum supremacy in near-term devices. *Nature Physics*, 14(6):595, 2018. DOI: 10.1038/s41567-018-0124-x 70, 85

[156] Adam Bouland, Bill Fefferman, Chinmay Nirkhe, and Umesh Vazirani.   Quantum supremacy and the complexity of random circuit sampling.   *ArXiv Preprint ArXiv:1803.04402*, 2018. 70

[157] Daniel Gottesman. Stabilizer codes and quantum error correction. *ArXiv Preprint quant-ph/9705052*, 1997. 77, 88, 141

[158] Rigetti computing. https://www.rigetti.com/systems 77, 83

[159] Pranav Gokhale, Yongshan Ding, Thomas Propson, Christopher Winkler, Nelson Leung, Yunong Shi, David I. Schuster, Henry Hoffmann, and Frederic T. Chong. Partial compilation of variational algorithms for noisy intermediate-scale quantum machines. In *Proc. of the 52nd Annual IEEE/ACM International Symposium on Microarchitecture*, pages 266–278, 2019. DOI: 10.1145/3352460.3358313 79, 131, 132, 140

[160] Ali Javadi-Abhari, Pranav Gokhale, Adam Holmes, Diana Franklin, Kenneth R. Brown, Margaret Martonosi, and Frederic T. Chong. Optimized surface code communication in superconducting quantum computers. In *Proc. of the 50th Annual IEEE/ACM International Symposium on Microarchitecture*, pages 692–705, 2017. DOI: 10.1145/3123939.3123949 79, 112

[161] Yongshan Ding, Adam Holmes, Ali Javadi-Abhari, Diana Franklin, Margaret Martonosi, and Frederic Chong. Magic-state functional units: Mapping and scheduling multi-level distillation circuits for fault-tolerant quantum architectures. In *51st Annual IEEE/ACM International Symposium on Microarchitecture (MICRO)*, pages 828–840, 2018. DOI: 10.1109/micro.2018.00072 79, 106, 113, 117, 118

[162] A. W. Cross. Unpublished. https://www.media.mit.edu/quanta/quanta-web/projects/qasm-tools/ 81

[163] Andrew W. Cross, Lev S. Bishop, John A. Smolin, and Jay M. Gambetta. Open quantum assembly language. *ArXiv Preprint ArXiv:1707.03429*, 2017. 82, 89

[164] S. Bourdeauducq, et al. Advanced Real-Time Infrastructure for Quantum physics, ARTIQ 1.0. zenodo. https://github.com/m-labs/artiq 82

[165] David C. McKay, Thomas Alexander, Luciano Bello, Michael J. Biercuk, Lev Bishop, Jiayin Chen, Jerry M. Chow, Antonio D. Córcoles, Daniel Egger, Stefan Filipp, et al. Qiskit backend specifications for openqasm and openpulse experiments. *ArXiv Preprint ArXiv:1809.03452*, 2018. 82, 106, 112, 113

[166] Alexander S. Green, Peter LeFanu Lumsdaine, Neil J. Ross, Peter Selinger, and Benoît Valiron. Quipper: A scalable quantum programming language. In *ACM SIGPLAN Notices*, 48:333–342, 2013. DOI: 10.1145/2491956.2462177 83, 89

[167] Andrei Lapets, Marcus P. da Silva, Mike Thome, Aaron Adler, Jacob Beal, and Martin Rötteler. Quafl: A typed DSL for quantum programming. In *Proc. of the 1st Annual Workshop on Functional Programming Concepts in Domain-Specific Languages*, pages 19–26, ACM, 2013. DOI: 10.1145/2505351.2505357 83

[168] Dave Wecker and Krysta M. Svore. Liqui|>: A software design architecture and domain-specific language for quantum computing. *ArXiv Preprint ArXiv:1402.4467*, 2014. 83

[169] Krysta M. Svore, Alan Geller, Matthias Troyer, John Azariah, Christopher Granade, Bettina Heim, Vadym Kliuchnikov, Mariia Mykhailova, Andres Paz, and Martin Roetteler. Q#: Enabling scalable quantum computing and development with a high-level domain-specific language. *ArXiv Preprint ArXiv:1803.00652*, 2018. 83, 89

[170] Ali JavadiAbhari, Shruti Patil, Daniel Kudrow, Jeff Heckey, Alexey Lvov, Frederic T. Chong, and Margaret Martonosi. ScaffCC: Scalable compilation and analysis of quantum programs. *Parallel Computing*, 45:2–17, 2015. DOI: 10.1016/j.parco.2014.12.001 83, 105, 113

[171] Damian S. Steiger, Thomas Häner, and Matthias Troyer. Projectq: An open source software framework for quantum computing. *Quantum*, 2(49):10–22331, 2018. DOI: 10.22331/q-2018-01-31-49 83

[172] Robert S. Smith, Michael J. Curtis, and William J. Zeng. A practical quantum instruction set architecture. *ArXiv Preprint ArXiv:1608.03355*, 2016. 83, 89

[173] Microsoft Quantum Computing. https://www.microsoft.com/en-us/quantum/ 83

[174] Amazon Braket. https://aws.amazon.com/braket/ 83

[175] Ashley Montanaro and Ronald de Wolf. A survey of quantum property testing. *ArXiv Preprint ArXiv:1310.2035*, 2013. 84, 86, 87

[176] Anne Broadbent. How to verify a quantum computation. *ArXiv Preprint ArXiv:1509.09180*, 2015. 84

[177] Urmila Mahadev. Classical verification of quantum computations. In *IEEE 59th Annual Symposium on Foundations of Computer Science (FOCS)*, pages 259–267, 2018. DOI: 10.1109/focs.2018.00033 84

[178] Ben W. Reichardt, Falk Unger, and Umesh Vazirani. Classical command of quantum systems. *Nature*, 496(7446):456, 2013. DOI: 10.1038/nature12035 84

[179] John Preskill. Quantum computing and the entanglement frontier. *ArXiv Preprint ArXiv:1203.5813*, 2012. 85

[180] Sergey Bravyi and David Gosset. Improved classical simulation of quantum circuits dominated by Clifford gates. *Physical Review Letters*, 116(25):250501, 2016. DOI: 10.1103/physrevlett.116.250501 85, 148, 164

[181] Harry Buhrman, Richard Cleve, John Watrous, and Ronald De Wolf. Quantum fingerprinting. *Physical Review Letters*, 87(16):167902, 2001. DOI: 10.1103/physrevlett.87.167902 87

[182] Hirotada Kobayashi, Keiji Matsumoto, and Tomoyuki Yamakami. Quantum Merlin-Arthur proof systems: Are multiple Merlins more helpful to Arthur? In *International Symposium on Algorithms and Computation*, pages 189–198, Springer, 2003. DOI: 10.1007/978-3-540-24587-2_21 87

[183] Masaru Kada, Harumichi Nishimura, and Tomoyuki Yamakami. The efficiency of quantum identity testing of multiple states. *Journal of Physics A: Mathematical and Theoretical*, 41(39):395309, 2008. DOI: 10.1088/1751-8113/41/39/395309 87

[184] Florian Mintert, Marek Kuś, and Andreas Buchleitner. Concurrence of mixed multipartite quantum states. *Physical Review Letters*, 95(26):260502, 2005. DOI: 10.1103/physrevlett.95.260502 87

[185] Aram W. Harrow and Ashley Montanaro. Testing product states, quantum merlin-arthur games and tensor optimization. *Journal of the ACM (JACM)*, 60(1):3, 2013. DOI: 10.1145/2432622.2432625 87

[186] Scott Aaronson and Daniel Gottesman. Identifying stabilizer states, 2008. 88

[187] David Perez-Garcia, Frank Verstraete, Michael M. Wolf, and J. Ignacio Cirac. Matrix product state representations. *ArXiv Preprint quant-ph/0608197*, 2006. 88

[188] Andrew M. Childs, Aram W. Harrow, and Paweł Wocjan. Weak fourier-schur sampling, the hidden subgroup problem, and the quantum collision problem. In *Annual Symposium on Theoretical Aspects of Computer Science*, pages 598–609, Springer, 2007. DOI: 10.1007/978-3-540-70918-3_51 88

[189] Otfried Gühne and Géza Tóth. Entanglement detection. *Physics Reports*, 474(1-6):1–75, 2009. DOI: 10.1016/j.physrep.2009.02.004 88

[190] Guoming Wang. Property testing of unitary operators. *Physical Review A*, 84(5):052328, 2011. DOI: 10.1103/physreva.84.052328 88, 89

[191] Man-Duen Choi. Completely positive linear maps on complex matrices. *Linear Algebra and its Applications*, 10(3):285–290, 1975. DOI: 10.1016/0024-3795(75)90075-0 88

[192] Andrzej Jamiołkowski. Linear transformations which preserve trace and positive semidefiniteness of operators. *Reports on Mathematical Physics*, 3(4):275–278, 1972. DOI: 10.1016/0034-4877(72)90011-0 88

[193] Ashley Montanaro and Tobias J. Osborne. Quantum boolean functions. *ArXiv Preprint ArXiv:0810.2435*, 2008. 89

[194] Lev Glebsky. Almost commuting matrices with respect to normalized Hilbert–Schmidt norm. *ArXiv Preprint ArXiv:1002.3082*, 2010. 89

[195] Robert Rand, Jennifer Paykin, and Steve Zdancewic. Qwire practice: Formal verification of quantum circuits in COQ. *ArXiv Preprint ArXiv:1803.00699*, 2018. DOI: 10.4204/eptcs.266.8 89

[196] Bruno Barras, Samuel Boutin, Cristina Cornes, Judicael Courant, Jean-Christophe Filliatre, Eduardo Gimenez, Hugo Herbelin, Gerard Huet, Cesar Munoz, Chetan Murthy, et al. The Coq proof assistant reference manual: Version 6.1. 1997. 89

[197] Matthew Amy. Towards large-scale functional verification of universal quantum circuits. *ArXiv Preprint ArXiv:1805.06908*, 2018. DOI: 10.4204/eptcs.287.1 89

[198] Mingsheng Ying. Floyd–hoare logic for quantum programs. *ACM Transactions on Programming Languages and Systems (TOPLAS)*, 33(6):19, 2011. DOI: 10.1145/2049706.2049708 89

[199] Mingsheng Ying. *Foundations of Quantum Programming*. Morgan Kaufmann, 2016. DOI: 10.1016/c2014-0-02660-3 89

[200] Dominique Unruh. Quantum relational hoare logic. *Proc. of the ACM on Programming Languages*, 3(POPL):33, 2019. DOI: 10.1145/3290346 89

[201] Matthew Amy, Martin Roetteler, and Krysta M. Svore. Verified compilation of space-efficient reversible circuits. In *International Conference on Computer Aided Verification*, pages 3–21, Springer, 2017. DOI: 10.1007/978-3-319-63390-9_1 89, 121

[202] André Van Tonder. A lambda calculus for quantum computation. *SIAM Journal on Computing*, 33(5):1109–1135, 2004. DOI: 10.1137/s0097539703432165 89

[203] Peter Selinger and Benoit Valiron. A lambda calculus for quantum computation with classical control. *Mathematical Structures in Computer Science*, 16(3):527–552, 2006. DOI: 10.1007/11417170_26 89

[204] Bob Coecke and Ross Duncan. Interacting quantum observables. In *International Colloquium on Automata, Languages, and Programming*, pages 298–310, Springer, 2008. DOI: 10.1007/978-3-540-70583-3_25 90

[205] Miriam Backens. The zx-calculus is complete for stabilizer quantum mechanics. *New Journal of Physics*, 16(9):093021, 2014. DOI: 10.1088/1367-2630/16/9/093021 90

[206] Amar Hadzihasanovic. The algebra of entanglement and the geometry of composition. *ArXiv Preprint ArXiv:1709.08086*, 2017. 90

[207] Ken Matsumoto and Kazuyuki Amano. Representation of quantum circuits with Clifford and pi/8 gates. *ArXiv Preprint ArXiv:0806.3834*, 2008. 94, 100, 101

[208] Neil J. Ross and Peter Selinger. Optimal ancilla-free clifford+ t approximation of z-rotations. *ArXiv Preprint ArXiv:1403.2975*, 2014. 94, 100, 101

[209] Simon Forest, David Gosset, Vadym Kliuchnikov, and David McKinnon. Exact synthesis of single-qubit unitaries over Clifford-cyclotomic gate sets. *Journal of Mathematical Physics*, 56(8):082201, 2015. DOI: 10.1063/1.4927100 94, 101

[210] Alex Bocharov, Yuri Gurevich, and Krysta M. Svore. Efficient decomposition of single-qubit gates into v basis circuits. *Physical Review A*, 88(1):012313, 2013. DOI: 10.1103/physreva.88.012313 94, 100, 101

[211] Vadym Kliuchnikov, Alex Bocharov, and Krysta M. Svore. Asymptotically optimal topological quantum compiling. *Physical Review Letters*, 112(14):140504, 2014. DOI: 10.1103/physrevlett.112.140504 94, 101, 105

[212] Alex Parent, Martin Roetteler, and Krysta M. Svore. Reversible circuit compilation with space constraints. *ArXiv Preprint ArXiv:1510.00377*, 2015. 99, 121, 122

[213] Charles Bennett. Time/space trade-offs for reversible computation. *SIAM Journal on Computing*, 18(4):766–776, 1989. DOI: 10.1137/0218053 99, 121, 122

[214] Aram W. Harrow, Avinatan Hassidim, and Seth Lloyd. Quantum algorithm for linear systems of equations. *Physical Review Letters*, 103(15):150502, 2009. DOI: 10.1103/physrevlett.103.150502 99, 121

[215] Vadym Kliuchnikov, Dmitri Maslov, and Michele Mosca. Fast and efficient exact synthesis of single qubit unitaries generated by Clifford and t gates. *ArXiv Preprint ArXiv:1206.5236*, 2012. 100, 149

[216] Brett Giles and Peter Selinger. Remarks on Matsumoto and Amano's normal form for single-qubit Clifford+ t operators. *ArXiv Preprint ArXiv:1312.6584*, 2013. 100, 101

[217] Matthew Amy, Andrew N. Glaudell, and Neil J. Ross. Number-theoretic characterizations of some restricted Clifford+ t circuits. *Quantum*, 4:252, 2020. DOI: 10.22331/q-2020-04-06-252 101

[218] Neil J. Ross. Optimal ancilla-free Clifford+ v approximation of z-rotations. *Quantum Information and Computation*, 15(11–12):932–950, 2015. 101

[219] Andrew N. Glaudell, Neil J. Ross, and Jacob M. Taylor. Optimal two-qubit circuits for universal fault-tolerant quantum computation. *ArXiv Preprint ArXiv:2001.05997*, 2020. 101

[220] Christopher M. Dawson and Michael A. Nielsen. The Solovay–Kitaev algorithm. *ArXiv Preprint quant-ph/0505030*, 2005. 103

[221] Matthew Amy, Dmitri Maslov, Michele Mosca, and Martin Roetteler. A meet-in-the-middle algorithm for fast synthesis of depth-optimal quantum circuits. *IEEE Transactions on Computer-Aided Design of Integrated Circuits and Systems*, 32(6):818–830, 2013. DOI: 10.1109/tcad.2013.2244643 103

[222] Aram W. Harrow, Benjamin Recht, and Isaac L. Chuang. Efficient discrete approximations of quantum gates. *Journal of Mathematical Physics*, 43(9):4445–4451, 2002. DOI: 10.1063/1.1495899 104

[223] Alex Bocharov, Shawn X. Cui, Martin Roetteler, and Krysta M. Svore. Improved quantum ternary arithmetics. *ArXiv Preprint ArXiv:1512.03824*, 2015. 105

[224] Alex Bocharov, Xingshan Cui, Vadym Kliuchnikov, and Zhenghan Wang. Efficient topological compilation for a weakly integral anyonic model. *Physical Review A*, 93(1):012313, 2016. DOI: 10.1103/physreva.93.012313 105

[225] Andrew N. Glaudell, Neil J. Ross, and Jacob M. Taylor. Canonical forms for single-qutrit Clifford+ t operators. *Annals of Physics*, 406:54–70, 2019. DOI: 10.1016/j.aop.2019.04.001 105

[226] Pranav Gokhale, Jonathan M. Baker, Casey Duckering, Natalie C. Brown, Kenneth R. Brown, and Frederic T. Chong. Asymptotic improvements to quantum circuits via qutrits. In *Proc. of the 46th International Symposium on Computer Architecture*, pages 554–566, ACM, 2019. DOI: 10.1145/3307650.3322253 105, 124

[227] Yongshan Ding, Xin-Chuan Wu, Adam Holmes, Ash Wiseth, Diana Franklin, Margaret Martonosi, and Frederic T. Chong. Square: Strategic quantum ancilla reuse for modular quantum programs via cost-effective uncomputation. *ArXiv Preprint ArXiv:2004.08539*, 2020. 106, 123

[228] Prakash Murali, Jonathan M. Baker, Ali Javadi-Abhari, Frederic T. Chong, and Margaret Martonosi. Noise-adaptive compiler mappings for noisy intermediate-scale quantum computers. In *Proc. of the 24th International Conference on Architectural Support for Programming Languages and Operating Systems*, pages 1015–1029, ACM, 2019. DOI: 10.1145/3297858.3304075 106, 140

[229] Swamit S. Tannu and Moinuddin K. Qureshi. Not all qubits are created equal: A case for variability-aware policies for NISQ era quantum computers. In *Proc. of the 24th International Conference on Architectural Support for Programming Languages and Operating Systems*, pages 987–999, ACM, 2019. DOI: 10.1145/3297858.3304007 106, 140

[230] Gushu Li, Yufei Ding, and Yuan Xie. Tackling the qubit mapping problem for NISQ era quantum devices. In *Proc. of the 24th International Conference on Architectural Support for Programming Languages and Operating Systems*, pages 1001–1014, 2019. DOI: 10.1145/3297858.3304023 106

[231] Shin Nishio, Yulu Pan, Takahiko Satoh, Hideharu Amano, and Rodney Van Meter. Extracting success from IBM's 20-qubit machines using error-aware compilation. *ArXiv Preprint ArXiv:1903.10963*, 2019. 106

[232] Prakash Murali, David C. McKay, Margaret Martonosi, and Ali Javadi-Abhari. Software mitigation of crosstalk on noisy intermediate-scale quantum computers. *ArXiv Preprint ArXiv:2001.02826*, 2020. DOI: 10.1145/3373376.3378477 106, 140, 141

[233] Daniel Gottesman and Isaac L. Chuang. Demonstrating the viability of universal quantum computation using teleportation and single-qubit operations. *Nature*, 402(6760):390–393, 1999. DOI: 10.1038/46503 106, 114, 148

[234] Daniel Gottesman and Isaac L. Chuang. Quantum teleportation is a universal computational primitive. *ArXiv Preprint quant-ph/9908010*, 1999. 106, 147

[235] Dong-Sheng Wang. Choi states, symmetry-based quantum gate teleportation, and stored-program quantum computing. *Physical Review A*, 101(5):052311, 2020. DOI: 10.1103/physreva.101.052311 106

[236] Mark Oskin, Frederic T. Chong, Isaac L. Chuang, and John Kubiatowicz. Building quantum wires: The long and the short of it. In *Proc. of the 30th Annual International Symposium on Computer Architecture*, pages 374–385, IEEE, 2003. DOI: 10.1145/871656.859661 106

[237] Sumeet Khatri, Ryan LaRose, Alexander Poremba, Lukasz Cincio, Andrew T. Sornborger, and Patrick J. Coles. Quantum-assisted quantum compiling. *Quantum*, 3:140, 2019. DOI: 10.22331/q-2019-05-13-140 106

[238] Gian Giacomo Guerreschi and Jongsoo Park. Two-step approach to scheduling quantum circuits. *Quantum Science and Technology*, 3(4):045003, 2018. DOI: 10.1088/2058-9565/aacf0b 110

[239] Jeff Heckey, Shruti Patil, Ali JavadiAbhari, Adam Holmes, Daniel Kudrow, Kenneth R. Brown, Diana Franklin, Frederic T. Chong, and Margaret Martonosi. Compiler management of communication and parallelism for quantum computation. In *ACM SIGARCH Computer Architecture News*, 43:445–456, 2015. DOI: 10.1145/2775054.2694357 112

[240] Adam Holmes, Yongshan Ding, Ali Javadi-Abhari, Diana Franklin, Margaret Martonosi, and Frederic T. Chong. Resource optimized quantum architectures for surface code implementations of magic-state distillation. *Microprocessors and Microsystems*, 67:56–70, 2019. DOI: 10.1016/j.micpro.2019.02.007 113

[241] Michael R. Garey and David S. Johnson. Crossing number is NP-complete. *SIAM Journal on Algebraic Discrete Methods*, 4(3):312–316, 1983. DOI: 10.1137/0604033 118

[242] Julia Chuzhoy, Yury Makarychev, and Anastasios Sidiropoulos. On graph crossing number and edge planarization. In *Proc. of the 22nd Annual ACM-SIAM Symposium on Discrete Algorithms*, pages 1050–1069, 2011. DOI: 10.1137/1.9781611973082.80 118

[243] Walter Schnyder. Embedding planar graphs on the grid. In *Proc. of the 1st Annual ACM-SIAM Symposium on Discrete Algorithms*, pages 138–148, Society for Industrial and Applied Mathematics, 1990. 118

[244] Brian W. Kernighan and Shen Lin. An efficient heuristic procedure for partitioning graphs. *The Bell System Technical Journal*, 49(2):291–307, 1970. DOI: 10.1002/j.1538-7305.1970.tb01770.x 118

[245] Earl R. Barnes. An algorithm for partitioning the nodes of a graph. *SIAM Journal on Algebraic Discrete Methods*, 3(4):541–550, 1982. DOI: 10.1109/cdc.1981.269534 118

[246] George Karypis and Vipin Kumar. Multilevel k-way hypergraph partitioning. *VLSI Design*, 11(3):285–300, 2000. DOI: 10.1145/309847.309954 118

[247] George Karypis and Vipin Kumar. Metis—unstructured graph partitioning and sparse matrix ordering system, version 2.0. 1995. 118

[248] François Pellegrini and Jean Roman. Scotch: A software package for static mapping by dual recursive bipartitioning of process and architecture graphs. In *International Conference on High-Performance Computing and Networking*, pages 493–498, Springer, 1996. DOI: 10.1007/3-540-61142-8_588 118

[249] Thomas M. J. Fruchterman and Edward M. Reingold. Graph drawing by force-directed placement. *Software: Practice and Experience*, 21(11):1129–1164, 1991. DOI: 10.1002/spe.4380211102 118

[250] Chun-Cheng Lin and Hsu-Chun Yen. A new force-directed graph drawing method based on edge—edge repulsion. *Journal of Visual Languages and Computing*, 23(1):29–42, 2012. DOI: 10.1109/iv.2005.10 118

[251] Yifan Hu. Efficient, high-quality force-directed graph drawing. *Mathematica Journal*, 10(1):37–71, 2005. 118

[252] William E. Donath and Alan J. Hoffman. Lower bounds for the partitioning of graphs. *IBM Journal of Research and Development*, 17(5):420–425, 1973. DOI: 10.1142/9789812796936_0044 119

[253] Michelle Girvan and Mark E. J. Newman. Community structure in social and biological networks. *Proc. of the National Academy of Sciences*, 99(12):7821–7826, 2002. DOI: 10.1073/pnas.122653799 119

[254] Miroslav Fiedler. Algebraic connectivity of graphs. *Czechoslovak Mathematical Journal*, 23(2):298–305, 1973. 119

[255] Barry D. Hughes. Random walks and random environments. 1995. 119

[256] Vincent D. Blondel, Jean-Loup Guillaume, Renaud Lambiotte, and Etienne Lefebvre. Fast unfolding of communities in large networks. *Journal of Statistical Mechanics: Theory and Experiment*, (10):10008, 2008. DOI: 10.1088/1742-5468/2008/10/p10008 119

[257] Jordi Duch and Alex Arenas. Community detection in complex networks using extremal optimization. *Physical Review E*, 72(2):027104, 2005. DOI: 10.1103/physreve.72.027104 119

[258] Tapas Kanungo, David M. Mount, Nathan S. Netanyahu, Christine D. Piatko, Ruth Silverman, and Angela Y. Wu. An efficient k-means clustering algorithm: Analysis and implementation. *IEEE Transactions on Pattern Analysis and Machine Intelligence*, 24(7):881–892, 2002. DOI: 10.1109/tpami.2002.1017616 119

[259] David Arthur and Sergei Vassilvitskii. k-means++: The advantages of careful seeding. In *Proc. of the 18th Annual ACM-SIAM Symposium on Discrete Algorithms*, pages 1027–1035, Society for Industrial and Applied Mathematics, 2007. 119

[260] Alexandru Paler, Austin G. Fowler, and Robert Wille. Faster manipulation of large quantum circuits using wire label reference diagrams. *Microprocessors and Microsystems*, 66:55–66, 2019. DOI: 10.1016/j.micpro.2019.02.008 119

[261] Alexandru Paler, Robert Wille, and Simon J. Devitt. Wire recycling for quantum circuit optimization. *Physical Review A*, 94(4):042337, 2016. DOI: 10.1103/physreva.94.042337 119

[262] Thomas Häner, Martin Roetteler, and Krysta M. Svore. Factoring using $2n + 2$ qubits with Toffoli based modular multiplication. *ArXiv Preprint ArXiv:1611.07995*, 2016. 120

[263] Craig Gidney. Factoring with $n + 2$ clean qubits and $n - 1$ dirty qubits. *ArXiv Preprint ArXiv:1706.07884*, 2017. 120

[264] Harry Buhrman, John Tromp, and Paul Vitányi. Time and space bounds for reversible simulation. In *International Colloquium on Automata, Languages, and Programming*, pages 1017–1027, Springer, 2001. DOI: 10.1088/0305-4470/34/35/308 121, 122

[265] Siu Man Chan, Massimo Lauria, Jakob Nordstrom, and Marc Vinyals. Hardness of approximation in pspace and separation results for pebble games. In *IEEE 56th Annual Symposium on Foundations of Computer Science*, pages 466–485, 2015. DOI: 10.1109/focs.2015.36 121

[266] Michael Patrick Frank and Thomas F. Knight Jr. Reversibility for efficient computing. Ph.D. thesis, Massachusetts Institute of Technology, Dept. of Electrical Engineering and Computer Science, 1999. 121

[267] Emanuel Knill. An analysis of Bennett's pebble game. *ArXiv Preprint math/9508218*, 1995. 121

[268] Balagopal Komarath, Jayalal Sarma, and Saurabh Sawlani. Pebbling meets coloring: Reversible pebble game on trees. *Journal of Computer and System Sciences*, 91:33–41, 2018. DOI: 10.1016/j.jcss.2017.07.009 121

[269] Giulia Meuli, Mathias Soeken, Martin Roetteler, Nikolaj Bjorner, and Giovanni De Micheli. Reversible pebbling game for quantum memory management. In *Design, Automation and Test in Europe Conference and Exhibition (DATE)*, pages 288–291, IEEE, 2019. DOI: 10.23919/date.2019.8715092 121

[270] Thomas Häner, Damian S. Steiger, Krysta Svore, and Matthias Troyer. A software methodology for compiling quantum programs. *Quantum Science and Technology*, 3(2):020501, 2018. DOI: 10.1088/2058-9565/aaa5cc 124

[271] Stephen S. Bullock, Dianne P. O'Leary, and Gavin K. Brennen. Asymptotically optimal quantum circuits for d-level systems. *Physical Review Letters*, 94(23):230502, 2005. DOI: 10.1103/physrevlett.94.230502 124

[272] Marek Perkowski, Anas Al-Rabadi, and Pawel Kerttopf. Multiple-valued quantum logic synthesis. 2002. 124

[273] S. S. Ivanov, H. S. Tonchev, and N. V. Vitanov. Time-efficient implementation of quantum search with qudits. *Physical Review A*, 85(6):062321, 2012. DOI: 10.1103/physreva.85.062321 124

[274] Alex Bocharov, Martin Roetteler, and Krysta M. Svore.   Factoring with qutrits: Shor's algorithm on ternary and metaplectic quantum architectures. *Physical Review A*, 96(1):012306, 2017. DOI: 10.1103/physreva.96.012306 124

[275] Greater Quantum Efficiency by Breaking Abstractions.   https://www.sigarch.org/greater-quantum-efficiency-by-breaking-abstractions/ 125

[276] Domenico d'Alessandro. *Introduction to Quantum Control and Dynamics*. Chapman and Hall/CRC, 2007. DOI: 10.1201/9781584888833 129

[277] Gabriel Turinici and Herschel Rabitz. Quantum wavefunction controllability. *Chemical Physics*, 267(1–3):1–9, 2001. DOI: 10.1016/s0301-0104(01)00216-6 129

[278] Gabriel Turinici and Herschel Rabitz.   Wavefunction controllability for finite-dimensional bilinear quantum systems. *Journal of Physics A: Mathematical and General*, 36(10):2565, 2003. DOI: 10.1088/0305-4470/36/10/316 129

[279] Warren S. Warren, Herschel Rabitz, and Mohammed Dahleh.   Coherent control of quantum dynamics: The dream is alive. *Science*, 259(5101):1581–1589, 1993. DOI: 10.1126/science.259.5101.1581 129

[280] Seth Lloyd. Coherent quantum feedback. *Physical Review A*, 62(2):022108, 2000. DOI: 10.1103/physreva.62.022108 129

[281] Moshe Shapiro and Paul Brumer. *Quantum Control of Molecular Processes*. John Wiley & Sons, 2012. DOI: 10.1002/9783527639700 129

[282] Mazyar Mirrahimi, Pierre Rouchon, and Gabriel Turinici.   Lyapunov control of bilinear Schrödinger equations.   *Automatica*, 41(11):1987–1994, 2005. DOI: 10.1016/j.automatica.2005.05.018 129

[283] Sen Kuang and Shuang Cong. Lyapunov control methods of closed quantum systems. *Automatica*, 44(1):98–108, 2008. DOI: 10.1016/j.automatica.2007.05.013 129

[284] Herschel Rabitz, Regina de Vivie-Riedle, Marcus Motzkus, and Karl Kompa. Whither the future of controlling quantum phenomena? *Science*, 288(5467):824–828, 2000. DOI: 10.1126/science.288.5467.824 129, 132

[285] Richard S. Judson and Herschel Rabitz. Teaching lasers to control molecules. *Physical Review Letters*, 68(10):1500, 1992. DOI: 10.1103/physrevlett.68.1500 129

[286] Marlan O. Scully and M. Suhail Zubairy.   Quantum optics, 1999.   DOI: 10.1017/cbo9780511813993 129

[287] Moshe Shapiro and Paul Brumer. Principles of the quantum control of molecular processes. *Principles of the Quantum Control of Molecular Processes*, by Moshe Shapiro, Paul
Brumer, page 250. Wiley-VCH, February, 2003. 129

[288] Andrew C. Doherty, Salman Habib, Kurt Jacobs, Hideo Mabuchi, and Sze M. Tan.
Quantum feedback control and classical control theory. *Physical Review A*, 62(1):012105,
2000. DOI: 10.1103/physreva.62.012105 129

[289] Daoyi Dong and Ian R. Petersen. Quantum control theory and applications: A survey. *IET Control Theory and Applications*, 4(12):2651–2671, 2010. DOI: 10.1049/iet-
cta.2009.0508 129

[290] Navin Khaneja, Timo Reiss, Cindie Kehlet, Thomas Schulte-Herbrüggen, and Steffen J.
Glaser. Optimal control of coupled spin dynamics: Design of NMR pulse sequences by
gradient ascent algorithms. *Journal of Magnetic Resonance*, 172(2):296–305, 2005. DOI:
10.1016/j.jmr.2004.11.004 130

[291] P. de Fouquieres, S. G. Schirmer, S. J. Glaser, and I. Kuprov. Second order gradient
ascent pulse engineering. *Journal of Magnetic Resonance*, 212:412–417, October 2011.
DOI: 10.1016/j.jmr.2011.07.023 130

[292] Nelson Leung, Mohamed Abdelhafez, Jens Koch, and David Schuster. Speedup for
quantum optimal control from automatic differentiation based on graphics processing
units. *Physical Review A*, 95(4):042318, 2017. DOI: 10.1103/physreva.95.042318 131,
132

[293] Mohamed Abdelhafez, David I. Schuster, and Jens Koch. Gradient-based optimal control of open quantum systems using quantum trajectories and automatic differentiation,
2019. DOI: 10.1103/physreva.99.052327 131

[294] Florian Dolde, Ville Bergholm, Ya Wang, Ingmar Jakobi, Boris Naydenov, Sébastien
Pezzagna, Jan Meijer, Fedor Jelezko, Philipp Neumann, Thomas Schulte-Herbrüggen,
Jacob Biamonte, and Jörg Wrachtrup. High-fidelity spin entanglement using
optimal control. *Nature Communications*, 5:3371 EP–, February 2014. DOI:
10.1038/ncomms4371 131

[295] Yi Chou, Shang-Yu Huang, and Hsi-Sheng Goan. Optimal control of fast and high-
fidelity quantum gates with electron and nuclear spins of a nitrogen-vacancy center in
diamond. *Phys. Rev. A*, 91:052315, May 2015. DOI: 10.1103/physreva.91.052315 131

[296] Steven Chu. Cold atoms and quantum control. *Nature*, 416(6877):206, 2002. DOI:
10.1038/416206a 132

[297] Howard M. Wiseman and Gerard J. Milburn. Quantum theory of optical feedback via homodyne detection. *Physical Review Letters*, 70(5):548, 1993. DOI: 10.1103/physrevlett.70.548 132

[298] Kenneth R. Brown, Aram W. Harrow, and Isaac L. Chuang. Arbitrarily accurate composite pulse sequences. *Physical Review A*, 70(5):052318, 2004. DOI: 10.1103/physreva.70.052318 132, 150

[299] Erwin L. Hahn. Spin echoes. *Physical Review*, 80(4):580, 1950. DOI: 10.1063/1.3066708 132

[300] Herman Y. Carr and Edward M. Purcell. Effects of diffusion on free precession in nuclear magnetic resonance experiments. *Physical Review*, 94(3):630, 1954. DOI: 10.1103/physrev.94.630 132

[301] Saul Meiboom and David Gill. Modified spin-echo method for measuring nuclear relaxation times. *Review of Scientific Instruments*, 29(8):688–691, 1958. DOI: 10.1063/1.1716296 132

[302] Jonas Bylander, Simon Gustavsson, Fei Yan, Fumiki Yoshihara, Khalil Harrabi, George Fitch, David G. Cory, Yasunobu Nakamura, Jaw-Shen Tsai, and William D. Oliver. Noise spectroscopy through dynamical decoupling with a superconducting flux qubit. *Nature Physics*, 7(7):565–570, 2011. DOI: 10.1038/nphys1994 132

[303] Mohamed Abdelhafez, David I. Schuster, and Jens Koch. Gradient-based optimal control of open quantum systems using quantum trajectories and automatic differentiation. *Physical Review A*, 99(5):052327, 2019. DOI: 10.1103/physreva.99.052327 132

[304] Navin Khaneja, Timo Reiss, Cindie Kehlet, Thomas Schulte-Herbrüggen, and Steffen J. Glaser. Optimal control of coupled spin dynamics: Design of NMR pulse sequences by gradient ascent algorithms. *Journal of Magnetic Resonance*, 172(2):296–305, 2005. DOI: 10.1016/j.jmr.2004.11.004 132

[305] Yunong Shi, Nelson Leung, Pranav Gokhale, Zane Rossi, David I. Schuster, Henry Hoffmann, and Frederic T. Chong. Optimized compilation of aggregated instructions for realistic quantum computers. In *Proc. of the 24th International Conference on Architectural Support for Programming Languages and Operating Systems*, pages 1031–1044, ACM, 2019. DOI: 10.1145/3297858.3304018 132, 140

[306] A. J. Skinner, M. E. Davenport, and Bruce E. Kane. Hydrogenic spin quantum computing in silicon: A digital approach. *Physical Review Letters*, 90(8):087901, 2003. DOI: 10.1103/physrevlett.90.087901 132

[307] Stephen H. Gunther, Frank Binns, Jack D. Pippin, Linda J. Rankin, Edward A. Burton, Douglas M. Carmean, and John M. Bauer. Methods and apparatus for thermal management of an integrated circuit die, September 7, 2004. U.S. Patent 6,789,037. 132

[308] Adam Holmes, Mohammad Reza Jokar, Ghasem Pasandi, Yongshan Ding, Massoud Pedram, and Frederic T. Chong. NISQ+: Boosting quantum computing power by approximating quantum error correction. *ArXiv Preprint ArXiv:2004.04794*, 2020. 132

[309] Swamit S. Tannu, Douglas M. Carmean, and Moinuddin K. Qureshi. Cryogenic-dram based memory system for scalable quantum computers: A feasibility study. In *Proc. of the International Symposium on Memory Systems*, pages 189–195, ACM, 2017. DOI: 10.1145/3132402.3132436 132

[310] Isaac L. Chuang and Michael A. Nielsen. Prescription for experimental determination of the dynamics of a quantum black box. *Journal of Modern Optics*, 44(11–12):2455–2467, 1997. DOI: 10.1080/09500349708231894 136

[311] J. F. Poyatos, J. Ignacio Cirac, and Peter Zoller. Complete characterization of a quantum process: The two-bit quantum gate. *Physical Review Letters*, 78(2):390, 1997. DOI: 10.1103/physrevlett.78.390 136

[312] Marcus P. da Silva, Olivier Landon-Cardinal, and David Poulin. Practical characterization of quantum devices without tomography. *Physical Review Letters*, 107(21):210404, 2011. DOI: 10.1103/physrevlett.107.210404 136

[313] Steven T. Flammia, David Gross, Yi-Kai Liu, and Jens Eisert. Quantum tomography via compressed sensing: Error bounds, sample complexity and efficient estimators. *New Journal of Physics*, 14(9):095022, 2012. DOI: 10.1088/1367-2630/14/9/095022 136

[314] Steven T. Flammia and Yi-Kai Liu. Direct fidelity estimation from few Pauli measurements. *Physical Review Letters*, 106(23):230501, 2011. DOI: 10.1103/physrevlett.106.230501 136

[315] Joseph Emerson, Robert Alicki, and Karol Życzkowski. Scalable noise estimation with random unitary operators. *Journal of Optics B: Quantum and Semiclassical Optics*, 7(10):S347, 2005. DOI: 10.1088/1464-4266/7/10/021 136

[316] Benjamin Lévi, Cecilia C. López, Joseph Emerson, and David G. Cory. Efficient error characterization in quantum information processing. *Physical Review A*, 75(2):022314, 2007. DOI: 10.1103/physreva.75.022314 136

[317] Emanuel Knill, Dietrich Leibfried, Rolf Reichle, Joe Britton, R. Brad Blakestad, John D. Jost, Chris Langer, Roee Ozeri, Signe Seidelin, and David J. Wineland. Randomized benchmarking of quantum gates. *Physical Review A*, 77(1):012307, 2008. DOI: 10.1103/physreva.77.012307 136

[318] Christoph Dankert, Richard Cleve, Joseph Emerson, and Etera Livine. Exact and approximate unitary 2-designs and their application to fidelity estimation. *Physical Review A*, 80(1):012304, 2009. DOI: 10.1103/physreva.80.012304 136

[319] Easwar Magesan, Jay M. Gambetta, and Joseph Emerson. Scalable and robust randomized benchmarking of quantum processes. *Physical Review Letters*, 106(18):180504, 2011. DOI: 10.1103/physrevlett.106.180504 136

[320] True-Q design by quantum benchmark. https://trueq.quantumbenchmark.com/index.html 137

[321] Easwar Magesan, Jay M. Gambetta, Blake R. Johnson, Colm A. Ryan, Jerry M. Chow, Seth T. Merkel, Marcus P. Da Silva, George A. Keefe, Mary B. Rothwell, Thomas A. Ohki, et al. Efficient measurement of quantum gate error by interleaved randomized benchmarking. *Physical Review Letters*, 109(8):080505, 2012. DOI: 10.1103/physrevlett.109.080505 137

[322] Joseph Emerson, Marcus Silva, Osama Moussa, Colm Ryan, Martin Laforest, Jonathan Baugh, David G. Cory, and Raymond Laflamme. Symmetrized characterization of noisy quantum processes. *Science*, 317(5846):1893–1896, 2007. DOI: 10.1126/science.1145699 137

[323] Joel Wallman, Chris Granade, Robin Harper, and Steven T. Flammia. Estimating the coherence of noise. *New Journal of Physics*, 17(11):113020, 2015. DOI: 10.1088/1367-2630/17/11/113020 137

[324] Joel J. Wallman, Marie Barnhill, and Joseph Emerson. Robust characterization of loss rates. *Physical Review Letters*, 115(6):060501, 2015. DOI: 10.1103/physrevlett.115.060501 137

[325] Andrew W. Cross, Easwar Magesan, Lev S. Bishop, John A. Smolin, and Jay M. Gambetta. Scalable randomised benchmarking of non-clifford gates. *NPJ Quantum Information*, 2:16012, 2016. DOI: 10.1038/npjqi.2016.12 137

[326] Arnaud Carignan-Dugas, Joel J. Wallman, and Joseph Emerson. Characterizing universal gate sets via dihedral benchmarking. *Physical Review A*, 92(6):060302, 2015. DOI: 10.1103/physreva.92.060302 137

[327] Bas Dirkse, Jonas Helsen, and Stephanie Wehner. Efficient unitarity randomized benchmarking of few-qubit Clifford gates. *Physical Review A*, 99(1):012315, 2019. DOI: 10.1103/physreva.99.012315 137

[328] Alexander Erhard, Joel James Wallman, Lukas Postler, Michael Meth, Roman Stricker, Esteban Adrian Martinez, Philipp Schindler, Thomas Monz, Joseph Emerson, and

Rainer Blatt. Characterizing large-scale quantum computers via cycle benchmarking. *ArXiv Preprint ArXiv:1902.08543*, 2019. DOI: 10.1038/s41467-019-13068-7 137, 150

[329] Joel J. Wallman and Joseph Emerson. Noise tailoring for scalable quantum computation via randomized compiling. *Physical Review A*, 94(5):052325, 2016. DOI: 10.1103/physreva.94.052325 139

[330] Prakash Murali, Norbert Matthias Linke, Margaret Martonosi, Ali Javadi Abhari, Nhung Hong Nguyen, and Cinthia Huerta Alderete. Full-stack, real-system quantum computer studies: Architectural comparisons and design insights. In *Proc. of the 46th International Symposium on Computer Architecture*, pages 527–540, ACM, 2019. DOI: 10.1145/3307650.3322273 140

[331] Markus Brink, Jerry M. Chow, Jared Hertzberg, Easwar Magesan, and Sami Rosenblatt. Device challenges for near term superconducting quantum processors: Frequency collisions. In *IEEE International Electron Devices Meeting (IEDM)*, pages 6–1, 2018. DOI: 10.1109/iedm.2018.8614500 140

[332] Gushu Li, Yufei Ding, and Yuan Xie. Towards efficient superconducting quantum processor architecture design. In *Proc. of the 25th International Conference on Architectural Support for Programming Languages and Operating Systems*, pages 1031–1045, 2020. DOI: 10.1145/3373376.3378500 140, 141

[333] M. D. Hutchings, Jared B. Hertzberg, Yebin Liu, Nicholas T. Bronn, George A. Keefe, Markus Brink, Jerry M. Chow, and B. L. T. Plourde. Tunable superconducting qubits with flux-independent coherence. *Physical Review Applied*, 8(4):044003, 2017. DOI: 10.1103/physrevapplied.8.044003 140

[334] R. Barends, C. M. Quintana, A. G. Petukhov, Yu Chen, D. Kafri, K. Kechedzhi, R. Collins, O. Naaman, S. Boixo, F. Arute, et al. Diabatic gates for frequency-tunable superconducting qubits. *Physical Review Letters*, 123(21):210501, 2019. DOI: 10.1103/physrevlett.123.210501 140

[335] R. Versluis, S. Poletto, N. Khammassi, B. Tarasinski, N. Haider, D. J. Michalak, A. Bruno, K. Bertels, and L. DiCarlo. Scalable quantum circuit and control for a superconducting surface code. *Physical Review Applied*, 8(3):034021, 2017. DOI: 10.1103/physrevapplied.8.034021 140

[336] Ferdinand Helmer, Matteo Mariantoni, Austin G. Fowler, Jan von Delft, Enrique Solano, and Florian Marquardt. Cavity grid for scalable quantum computation with superconducting circuits. *EPL (Europhysics Letters)*, 85(5):50007, 2009. DOI: 10.1209/0295-5075/85/50007 140

[337] Yu Chen, C. Neill, P. Roushan, N. Leung, M. Fang, R. Barends, J. Kelly, B. Campbell, Z. Chen, B. Chiaro, et al. Qubit architecture with high coherence and fast tunable coupling. *Physical Review Letters*, 113(22):220502, 2014. DOI: 10.1103/physrevlett.113.220502 140

[338] A Robert Calderbank and Peter W. Shor. Good quantum error-correcting codes exist. *Physical Review A*, 54(2):1098, 1996. DOI: 10.1103/physreva.54.1098 141

[339] Andrew M. Steane. Simple quantum error-correcting codes. *Physical Review A*, 54(6):4741, 1996. DOI: 10.1103/physreva.54.4741 141

[340] Aleksei Yur'evich Kitaev. Quantum computations: Algorithms and error correction. *Uspekhi Matematicheskikh Nauk*, 52(6):53–112, 1997. DOI: 10.1070/rm1997v052n06abeh002155 141

[341] Daniel Gottesman. An introduction to quantum error correction. In *Proc. of Symposia in Applied Mathematics*, 58:221–236, 2002. DOI: 10.1090/psapm/058/1922900 141

[342] Emanuel Knill. Quantum computing with realistically noisy devices. *Nature*, 434(7029):39–44, 2005. DOI: 10.1038/nature03350 147, 148

[343] Swamit S. Tannu, Zachary A. Myers, Prashant J. Nair, Douglas M. Carmean, and Moinuddin K. Qureshi. Taming the instruction bandwidth of quantum computers via hardware-managed error correction. In *Proc. of the 50th Annual IEEE/ACM International Symposium on Microarchitecture*, pages 679–691, 2017. DOI: 10.1145/3123939.3123940 148

[344] Nemanja Isailovic, Mark Whitney, Yatish Patel, and John Kubiatowicz. Running a quantum circuit at the speed of data. In *ACM SIGARCH Computer Architecture News*, 36:177–188, IEEE Computer Society, 2008. DOI: 10.1109/isca.2008.5 148

[345] Dave Wecker, Bela Bauer, Bryan K. Clark, Matthew B. Hastings, and Matthias Troyer. Gate-count estimates for performing quantum chemistry on small quantum computers. *Physical Review A*, 90(2):022305, 2014. DOI: 10.1103/physreva.90.022305 149

[346] Sergey Bravyi and Jeongwan Haah. Magic-state distillation with low overhead. *Physical Review A*, 86(5):052329, 2012. DOI: 10.1103/physreva.86.052329 149

[347] Lorenza Viola, Emanuel Knill, and Seth Lloyd. Dynamical decoupling of open quantum systems. *Physical Review Letters*, 82(12):2417, 1999. DOI: 10.1103/physrevlett.82.2417 150

[348] Robin Harper, Steven T. Flammia, and Joel J. Wallman. Efficient learning of quantum noise. *ArXiv Preprint ArXiv:1907.13022*, 2019. 150

[349] Aram W. Harrow and Michael A. Nielsen. Robustness of quantum gates in the presence of noise. *Physical Review A*, 68(1):012308, 2003. DOI: 10.1103/physreva.68.012308 150

[350] Mauricio Gutiérrez, Lukas Svec, Alexander Vargo, and Kenneth R. Brown. Approximation of realistic errors by Clifford channels and Pauli measurements. *Physical Review A*, 87(3):030302, 2013. DOI: 10.1103/physreva.87.030302 150, 164

[351] Daniel Puzzuoli, Christopher Granade, Holger Haas, Ben Criger, Easwar Magesan, and David G. Cory. Tractable simulation of error correction with honest approximations to realistic fault models. *Physical Review A*, 89(2):022306, 2014. DOI: 10.1103/physreva.89.022306 150

[352] Dorit Aharonov and Michael Ben-Or. Fault-tolerant quantum computation with constant error rate. *SIAM Journal on Computing*, 2008. DOI: 10.1137/s0097539799359385 150

[353] M. Nest. Classical simulation of quantum computation, the Gottesman–Knill theorem, and slightly beyond. *ArXiv Preprint ArXiv:0811.0898*, 2008. 152

[354] Alison L. Gibbs and Francis Edward Su. On choosing and bounding probability metrics. *International Statistical Review*, 70(3):419–435, 2002. DOI: 10.1111/j.1751-5823.2002.tb00178.x 153

[355] Costin Bădescu, Ryan O'Donnell, and John Wright. Quantum state certification. In *Proc. of the 51st Annual ACM SIGACT Symposium on Theory of Computing*, pages 503–514, 2019. DOI: 10.1145/3313276.3316344 153

[356] Xin-Chuan Wu, Sheng Di, Franck Cappello, Hal Finkel, Yuri Alexeev, and Frederic T. Chong. Amplitude-aware lossy compression for quantum circuit simulation. *ArXiv Preprint ArXiv:1811.05140*, 2018. 155

[357] Xin-Chuan Wu, Sheng Di, Franck Cappello, Hal Finkel, Yuri Alexeev, and Frederic T. Chong. Memory-efficient quantum circuit simulation by using lossy data compression. *ArXiv Preprint ArXiv:1811.05630*, 2018. 155

[358] Richard Jozsa and Noah Linden. On the role of entanglement in quantum-computational speed-up. In *Proc. of the Royal Society of London A: Mathematical, Physical and Engineering Sciences*, 459:2011–2032, The Royal Society, 2003. DOI: 10.1098/rspa.2002.1097 156

[359] Guifré Vidal. Efficient classical simulation of slightly entangled quantum computations. *Physical Review Letters*, 91(14):147902, 2003. DOI: 10.1103/physrevlett.91.147902 156

[360] Richard Jozsa. On the simulation of quantum circuits. *ArXiv Preprint quant–ph/0603163*, 2006. 156

[361] Maarten Van den Nest, Wolfgang Dür, Guifré Vidal, and H. J. Briegel. Classical simulation versus universality in measurement-based quantum computation. *Physical Review A*, 75(1):012337, 2007. DOI: 10.1103/physreva.75.012337 156

[362] Daniel Gottesman. Talk at international conference on group theoretic methods in physics. *arXiv preprint quant-ph/9807006*, 1998. 158

[363] Scott Aaronson and Daniel Gottesman. Improved simulation of stabilizer circuits. *Physical Review A*, 70(5):052328, 2004. DOI: 10.1103/physreva.70.052328 159

[364] Igor L. Markov and Yaoyun Shi. Simulating quantum computation by contracting tensor networks. *SIAM Journal on Computing*, 38(3):963–981, 2008. DOI: 10.1137/050644756 162

[365] Sergio Boixo, Sergei V. Isakov, Vadim N. Smelyanskiy, and Hartmut Neven. Simulation of low-depth quantum circuits as complex undirected graphical models. *ArXiv Preprint ArXiv:1712.05384*, 2017. 163

[366] Mauricio Gutiérrez and Kenneth R. Brown. Comparison of a quantum error-correction threshold for exact and approximate errors. *Physical Review A*, 91(2):022335, 2015. DOI: 10.1103/physreva.91.022335 164

[367] Hakop Pashayan, Joel J. Wallman, and Stephen D. Bartlett. Estimating outcome probabilities of quantum circuits using quasiprobabilities. *Physical Review Letters*, 115(7):070501, 2015. DOI: 10.1103/physrevlett.115.070501 164

[368] Mark Howard and Earl Campbell. Application of a resource theory for magic states to fault-tolerant quantum computing. *Physical Review Letters*, 118(9):090501, 2017. DOI: 10.1103/physrevlett.118.090501 164

[369] Sergey Bravyi, Graeme Smith, and John A. Smolin. Trading classical and quantum computational resources. *Physical Review X*, 6(2):021043, 2016. DOI: 10.1103/physrevx.6.021043 164

[370] Eliot Kapit. Hardware-efficient and fully autonomous quantum error correction in superconducting circuits. *Physical Review Letters*, 116(15):150501, 2016. DOI: 10.1103/physrevlett.116.150501 166

[371] Kanav Setia, Sergey Bravyi, Antonio Mezzacapo, and James D. Whitfield. Superfast encodings for fermionic quantum simulation. *ArXiv Preprint ArXiv:1810.05274*, 2018. DOI: 10.1103/physrevresearch.1.033033 166

Authors' Biographies

YONGSHAN DING

Yongshan Ding is a fourth-year graduate student in the Department of Computer Science at the University of Chicago, advised by Fred Chong. Before UChicago, he received his dual B.Sc. degrees in Computer Science and Physics from Carnegie Mellon University. His research interests are in computer architectures, quantum algorithms, quantum information, and error correction. He builds systems that enable efficient scheduling and mapping from high-level circuits to noisy near-term devices and designs algorithms using tools from property testing, representation theory, and topology.

FREDERIC T. CHONG

Fred Chong is the Seymour Goodman Professor in the Department of Computer Science at the University of Chicago. He is also Lead Principal Investigator for the EPiQC Project (Enabling Practical-scale Quantum Computing), an NSF Expedition in Computing. Chong received his Ph.D. from MIT in 1996 and was a faculty member and Chancellor's fellow at UC Davis from 1997–2005. He was also a Professor of Computer Science, Director of Computer Engineering, and Director of the Greenscale Center for Energy-Efficient Computing at UCSB from 2005–2015. He is a recipient of the NSF CAREER award, the Intel Outstanding Researcher Award, and eight best paper awards. His research interests include emerging technologies for computing, quantum computing, multicore and embedded architectures, computer security, and sustainable computing.